T0132582

Borrowed Knowledge

Borrowed Knowledge: Chaos Theory and the Challenge of Learning across Disciplines

Stephen H. Kellert

The University of Chicago Press :: Chicago and London

Stephen H. Kellert is professor of philosophy at Hamline University in
St. Paul, Minnesota.

The University of Chicago Press, Chicago 60637
The University of Chicago Press, Ltd., London
© 2008 by The University of Chicago
All rights reserved. Published 2008
Printed in the United States of America

17 16 15 14 13 12 11 10 09 08   1 2 3 4 5

ISBN-13: 978-0-226-42978-6 (cloth)
ISBN-10: 0-226-42978-4 (cloth)

Library of Congress Cataloging-in-Publication Data

Kellert, Stephen H.
    Borrowed knowledge : chaos theory and the challenge of
learning across disciplines / Stephen H. Kellert.
        p. cm.
    Includes bibliographical references and index.
    ISBN-13: 978-0-226-42978-6 (cloth : alk. paper)
    ISBN-10: 0-226-42978-4 (cloth : alk. paper)
    1. Chaotic behavior in systems. 2. Science—Philosophy.
3. Interdisciplinary approach to knowledge. I. Title.
Q172.5.C45K449   2008
003′.857—dc22
                                    2008011080

To Sara,   *who was right all along*

Contents

Acknowledgments

This project would never have gotten started without the advice and encouragement of the late Susan Abrams, and it would never have gotten finished without the wisdom, support, and compassionate prodding supplied by Jim Bonilla. I owe a special debt to all my teachers, and especially to the late Donald Summersdorf, who gave me a copy of Samuelson's *Economics* to keep me busy, and to Diane Rubinstein, who first introduced me to poststructuralist thought.

Crucial encouragement along the way has come from Ron Giere, Karen Warren, Janet Folina, Arthur Fine, Margaret McLaren, Alice Dreger, Stephen Parker, Terry Kent, and the remarkable Mike Reynolds. Many thanks are due to those who read chapters of this work at various stages: Rose Mack, Mark Olson, Nancy Holland, Alan Gross, Helen Longino, Ron Giere, Susan Williams, Maxine Eichner, Jordan Macknick, Alan Silva, Matt Ryg, and Sara Mack. I am especially grateful to Mike Reynolds, Duane Cady, Cally Anderson, Andrew Abruzzese, and Terry Kent, who read the entire manuscript and gave helpful suggestions on numerous points. Conversations with Sam Imbo, Dave Doyle, Ken Waters, Jeff Turner, Paul Jessup, and Sharon Preves have all helped improve this work, and I have also benefited greatly from the responses of two reviewers and from opportunities to present portions of this work to audiences at Vassar College, the University of Minnesota, Macalester College, Miami University, and the Minnesota workshop on pluralism.

I am grateful to Hamline University for support in the form of Hanna grants for summer research and a sabbatical, and I am especially grateful to my departmental colleagues Duane Cady, Nancy Holland, and Sam Imbo for both their congeniality and their flexibility in scheduling classes so as to allow me time for research. Thanks to Christie Henry and Kathryn Gohl, and the staff of the University of Chicago Press for all their help. Many thanks also to my families, the Kellerts and the Macks, for their support. And most of all, thanks are due to Sara L. Mack, my beloved partner in life.

Some early pieces of this project appeared, in very different form, in the journals *Philosophy of Science, Configurations,* and *Philosophy, Psychiatry & Psychology,* and in the volume *Chaos and Society,* edited by A. Albert. Part of chapter 2 appeared as "Disciplinary Pluralism for Science Studies" in the volume *Scientific Pluralism,* copyright 2006 by the Regents of the University of Minnesota; I thank the University of Minnesota Press for permission to include it here.

1

What Was Chaos Theory, and Why Would People Want to Borrow It?

The Phenomenon

Each time I taught a class on interpretations and applications of chaos theory, I asked the students, for their final project, to find out how this recent work had been used outside of its primary home in the natural sciences. "Pick an area of study that interests you and find an article or two about the implications of chaotic dynamics for that field," I said, hoping that they would not come up empty-handed. They never did. The student interested in theology found a number of articles on chaos theory as a model of divine intervention, and the student interested in family therapy found articles on the nonlinear dynamics of sibling relationships. Fashion marketing, anthropology, music—it seemed as if no subject failed to capitalize on this new field.

How widespread is this phenomenon? A poetic convention at least as old as the Hebrew Psalms seeks to convey a sense of completeness or omnipresence by use of an acrostic list from A to Z.[1] Consider that chaos theory has been used in academic studies of

- art (Shearer 1992)
- business strategy (Bechtold 1997)

- curriculum development (Iannone 1995)
- Daoism (Jones and Culliney 1999)
- environmental awareness (Richards 2001)
- forensic sciences (Kirby 1990)
- globalization (Appadurai 1996)
- historical explanation (Dyke 1990)
- incest (Pepinsky 1997)
- Jungian theories of the self (Butz 1992)
- Korean stock market prices (Sewell et al. 1996)
- leadership (Wheatley 1992)
- memory (Tsuda 2001)
- nursing (Haigh 2002)
- the Other (Martin-Vallas 2005)
- planning (Cartwright 1991)
- *Don Quixote* (Flores 2002)
- race (Ernest 2006)
- social work (Hudson 2000)
- theology (Russell, Murphy, and Peacocke 1995)
- urban economic development (Howells 1997)
- vaccination policies (Piccardi and Lazzaris 1998)
- war (Saperstein 1984)
- *The X-Files* (Wildermuth 1999)
- Yellow River flood cycles (Zhou, Ma, and Wang 2002)
- zero-sum games (Sato, Akiyama, and Farmer 2002)

Some of these attempts to make use of chaos theory merit serious attention, some are harmless speculation, and some are simply misleading. This phenomenon of "borrowing" from the natural sciences—the attempt to transfer concepts, methods, and results to other disciplines—offers a case study that can illuminate important features of the way we create knowledge. For as we confront the phenomenon of borrowed knowledge, a number of questions arise. For example, why do people borrow knowledge, and how do they go about it? What do they hope to accomplish by borrowing knowledge, and what do they actually accomplish? When does it work well, and when does it work poorly? In what follows, I address these questions by surveying the uses of chaos theory in three particular fields: economics, legal theory, and literary studies. This work provides a theoretical vocabulary and a set of critical frameworks for the rigorous examination of borrowing while also engaging with recent debates about interdisciplinary research and the relationship between the sciences and the humanities. In addition to sketching some of the problems and mistakes encountered in borrowing, I argue for the possibility of the fruitful use of chaotic dynamics in work outside the natural sciences. This very possibility stands in sharp contrast with some influential conceptions of knowledge that would dismiss even the illuminating instances of borrowing as irrelevant or illegitimate. So the details of particular cross-disciplinary appropriations are less important than the challenges that borrowing poses for our views about how knowledge gets made. My goal is not to develop a comprehensive catalog of uses and abuses, nor is it to develop universal rules for the utilization of physics by researchers outside of the sciences. Instead, I seek to identify useful patterns and cautions for the transfer of scientific knowledge across disciplines, with the ultimate purpose of discerning lessons for our knowledge-making pursuits—lessons that go beyond the currently fashionable science of nonlinear dynamics. Cross-disciplinary borrowing did not begin with chaos theory, and it will not end when the glamour finally passes from this latest fad.

I approach this enterprise from my own academic home in the discipline of philosophy, and specifically the philosophy of science. Scientific knowledge receives a great deal of respect for its power to help us understand and cope with the natural world. Partly because of this power, philosophers have expended much effort in examining it, usually by looking at the production of scientific knowledge and the character of the knowledge produced. Some have sought to uncover the special features of scientific investigation that enable us to gain understanding, truth, predictive power, technical skill, or some other favorite measure of epistemic success. Others have focused on our current best scientific theories, seeking to resolve conceptual riddles or elicit metaphysical lessons. In what follows, I am concerned not with the production of scientific knowledge but with its consumption by researchers in other disciplines. In my previous book I undertook to clarify the nature and implications of our knowledge of chaotic dynamics, but here I explore the ways in which that knowledge gets borrowed in order to make new knowledge in different fields. As a consequence, this book involves a philosophical examination of the uses of science, but it is not "philosophy of science" in the traditional sense characterized earlier. Nonetheless, readers from other disciplines will discern my philosophical training in my concern for specifying the meanings of particular terms, in my sustained focus on the strength of arguments, and in my insistence on the need to evaluate as well as describe.

Philosophers must balance the two sides of our professional training. On the one hand, philosophy has inherited the tradition of speculation (although scientists are often more willing to engage in speculation than philosophers these days). Speculation includes many activities, one of which is the attempt to make connections among widely disparate fields of human experience. This aspect of philosophical training makes me sympathetic to those who search for a bold new synthesis of different areas of knowledge. On the other hand, contemporary philosophy in the English language has been strongly identified with conceptual analysis and criticism of arguments. It is this philosophical tradition that makes me highly suspicious of casually flexible uses of language or suggestive but tenuous leaps of reasoning such as are found in some appropriations of chaos theory. Although the field of philosophy of science has played an important role in discussions of the nature of science, philosophers have no special authority to decide what counts as "uses" or "misuses" of chaos theory. Still, philosophical analysis can address the important task of critically evaluating the interpretations, extensions, and appropriations of chaotic dynamics. Such analysis can help to sort out the broad conceptual implications of scientific results and to clarify possible connections among fields of inquiry.

In what follows, I employ both of these philosophical roles to examine some of the ways researchers have drawn relationships between chaos theory and other fields of inquiry. Although focusing on three specific points of contact between the natural sciences and other disciplines, this analysis is directed at a wider discussion regarding the ways knowledge is and ought to be translated across disciplines. In the remainder of this introductory chapter, I spell out what my project entails as well as what I do not address. I begin by setting out some clarifications of chaos theory and borrowed knowledge.

What Chaos Theory Was

The term "chaos theory" appears primarily in popularized versions of science (Hayles 1990, 8; Weingart and Maasen 1997, 466). Most scientists no longer use the term because in important respects, chaos theory is not about chaos and is not even a theory. In the first place, the "chaos" studied by nonlinear dynamics should not be understood as a primal welter of utterly unintelligible randomness (contra Slethaug 2000, 131; see also Assad 1991). And in the second place, chaos theory does not provide anything like the tidy logical structure of Newton's theory of universal gravitation or the sweeping explanatory power of the germ theory of disease. Even someone as closely associated with chaos theory as Ilya Prigogine has pointed out that chaos is just one kind of behavior examined by the more fundamental method of nonlinear dynamics. According to Prigogine, chaos theory is the flashy sideshow that nonetheless plays an important role in the circus (Day and Chen 1993, 324).

What, then, is chaos theory? It is helpful to consider the types of things that have served as standard examples for physical systems. First, Newtonian physics excelled at studying systems with stable behavior that regularly repeated—systems that behave like clocks. Clocklike behavior is simple because it is periodic and predictable, and it usually arises from simple systems with a few parts obeying strict mathematical laws: simple system, simple behavior. The physics of the eighteenth century focused on clocks and tended to see much (if not all) of the universe as clocklike (see Kellert 1993, chap. 5). In the nineteenth century, physics expanded to consider a second kind of system, which can be represented by the iconic image of the pressure cooker. A pressure cooker has trillions of molecules inside it, particles of air and water vapor bouncing around in an incomprehensibly complicated way. Although each individual particle may behave in a clocklike way on its own, it suffers so many collisions with its fellow particles that to predict its exact path would be practically impossible: complicated system, complicated behavior. Yet we can understand the behavior of the system as a whole in terms of its average properties: when you turn up the heat, the pressure will increase, and if you were to compress the cooker the temperature would go up. Physics can study general properties of a system such as temperature and pressure in terms of the average behavior of all its subsystems by using a probabilistic approach.

Chaotic systems are unlike both clocks and pressure cookers. Consider a taffy-pulling machine: it is a simple device that subjects the blob of taffy to a strictly defined manipulation—no need to consider trillions of interacting subsystems here. But if you put a small drop of food coloring in the taffy, the machine will soon stretch and fold the candy so that the color is spread throughout the taffy.[2] Two particles of dye that started out near each other might well end up on opposite sides of the big blob, and it is difficult if not impossible to predict where they will be. A chaotic system obeys simple rules like a clock but displays complicated and unpredictable behavior like a pressure cooker. Chaos theory offers a new kind of model physical system: simple system, complicated behavior.

In brief, chaos theory is the study of unpredictable behavior in simple, bounded, deterministic systems. Such behavior is extremely complicated because it never repeats, and it is unpredictable because of its celebrated sensitive dependence on initial conditions: even extremely small amounts of vagueness in specifying where the system starts render one utterly unable to predict where the system will end up. This sensitivity is often characterized in terms of the so-called butterfly effect, according to which our inability to know about the flapping of a butterfly's wings in some faraway place leads to our inability to correctly predict momentous weather phenomena in our own backyard (Gleick 1987, 11–31). The unpredictability of chaotic systems should not be thought of as a temporary inconvenience; instead, the unpredictability represents an in-principle limitation on the ability to make exact predictions about the specific details of the system's behavior (see Kellert 1993, chap. 2).

Chaotic behavior occurs only in *nonlinear* systems, that is, systems in which the relevant variables are not related one to another according to strict proportionality.[3] And chaotic behavior is investigated using the mathematical tools that go by the name "dynamical systems theory" or just "dynamics."[4] Thus, chaos theory should be understood as a subfield of *nonlinear dynamics*: all chaotic dynamics are necessarily nonlinear, but not all nonlinear dynamics exhibit chaos. Scholars who borrow from chaos theory sometimes use elements from the broader field of nonlinear dynamics, so I occasionally seem to switch between referring to "chaos theory" and "nonlinear dynamics" as if the terms name the same thing, but they do not. As a subcategory of nonlinear dynamics, chaos theory is especially noted for its investigation of patterns, such as "strange attractors," found within seemingly random data sets. One of the most famous of these is the Lorenz attractor, which arises from a highly simplified mathematical model that meteorologist Edward Lorenz developed for studying convection in the atmosphere. For some values of the parameters involved, this attractor looks like two thin disks spliced together; as the system evolves in time, the mathematical point representing the state of the system spirals around one disk for a while and then switches abruptly and unpredictably to swirl around the other. This shape, like all strange attractors, is an abstract geometrical entity representing the long-term behavior of a system in visual form. Because the Lorenz attractor exists only in mathematical "phase space," which is very different from the space in which you and I live, the only "attraction" it exerts is a result of the fact that, over time, the path representing the behavior of the system comes to closely resemble the attractor.

A cross-sectional view through a "slice" of a strange attractors can display an astonishingly intricate pattern in which each individual stripe or layer appears, on closer examination, to manifest the overall pattern on a smaller scale. This feature of "self-similarity" is characteristic of the curious geometrical objects known as fractals, so chaos theory has become closely associated with the fractal mathematics promoted by Benoit Mandelbrot. We can see a living approximation of a fractal shape by going to the market and taking a look at the vegetable known as romanesco broccoli. It looks like a bumpy green cone, and each of its big bumps is itself shaped like a bumpy green cone. And each of the smaller bumps on the big bumps is shaped like a bumpy green cone. Fractals are like that, only the self-similarity repeats at smaller and smaller scales, all the way down to the infinitesimal. An important quantitative measure of fractal structure is called the "dimension" of the attractor, which (roughly speaking) corresponds to how tightly the layers are packed together, or how much room is taken up by the bumps. Because this measurement can take on values other than just 1, 2, 3, etc., fractals are sometimes said to have a "fractional dimension."

Many scientists investigating complicated behavior in physical systems have begun to use the techniques of nonlinear dynamics and fractal mathematics. Typically, an experimental situation will yield a long time series of measurements, which investigators subject to a variety of techniques in order to understand the long-term and large-scale behavior of the system. And some mathematical models used to study these systems display chaotic dynamics: they are governed by strictly deterministic laws, yet their behavior is so complicated as to be unpredictable. Small tabletop physical systems were the first to be subjected to many of these techniques in the 1980s, including the celebrated experiment that reconstructed a chaotic attractor based on measurements of the time between drips from a faucet. Since then, controlled chemical reactions and vibrating metal beams have provided additional experimental situations well suited for study with dynamical techniques.

Systems outside the physical sciences present a genuine challenge to such attempts, though not an impossible one. In epidemiology and ecology, for instance, systems are much harder to control, making the discovery and analysis of dynamical chaos within them more controversial. In psychology, the work of Walter Freeman demonstrates the possibility of fruitful applications of the techniques of nonlinear dynamics within neurophysiology (Skarda and Freeman 1991). When quantitative techniques cannot be applied, chaos theory may be used as a source of metaphors. Although there are numerous instances of, for example, legal scholars drawing metaphorical conclusions from natural selection (Roe 1996; Clark 1981) or quantum mechanics (Tribe 1989; Veilleux 1987; Porter 1991), I focus on the rich array of interdisciplinary borrowings that arose from the vogue for chaos theory that peaked in the 1990s.

What Chaos Theory Is Not

One way to understand chaos theory more fully is to consider what it is not. For example, the unpredictability of chaotic behavior does not rest on quantum-mechanical uncertainty or indeterminacy (contra Slethaug 2000, xi; Hawkins 1995, 95), and in fact chaotic systems scrupulously obey the deterministic laws of old-fashioned Newtonian physics. Thus, chaos theory should not be considered a "deviant science" that will replace Newton's laws (contra Brion 1995, 181; see also Demastes 1998, 75; Gross and Levitt 1994, 98).[5] Chaos is not the same as entropy, which is a mathematical measure of the disorderliness in large pressure cooker–type systems, so chaos theory should be distinguished from the science of thermodynamics that studies such systems (contra Moran 1997, 174). Neither should it be conflated with "complexity theory," which studies the intricate behavior that can result from large numbers of subsystems interacting according to simple rules (contra Slethaug 2000, xii, xv). Many scholars who borrowed from chaos theory in the 1990s now draw ideas from a broader field called "complexity science." It may be that quantum mechanics, thermodynamics, complexity theory, and strange attractors are all part of this new scientific superparadigm that reexamines randomness. Perhaps dissipative structures, self-organization, and autopoeisis are also part of the mix. But for the purposes of this book, I focus more narrowly.[6]

Because of this narrower focus I will not be considering the work of Ilya Prigogine, a Nobel Prize–winning chemist whose work has also inspired a great deal of cross-disciplinary borrowing. Although his work is often treated as part of chaos theory in a broad sense, it actually has few connections to research on strange attractors and the like (Hayles 1990, 10; Matheson and Kirchhoff 1997, 32; but see Hayles 1991, 13). Prigogine's investigation of "dissipative structures" deals with systems that magnify small fluctuations into "self-organizing" large-scale order, but the stable patterns that emerge in self-organization are structures in actual space, unlike strange attractors that exist only in abstract mathematical representations (contra Porush 1991, 72). This difference gives his work quite a different philosophical spin. Compare the title of a standard textbook on chaos theory and strange attractors—*Order within Chaos*—and that of a book outlining Prigogine's conception of our "new dialogue with Nature," *Order out of Chaos*.

Prigogine's work also looks at the large-scale changes brought about by amplification of microscopic fluctuations, but this phenomenon is not deterministic chaos. The conflation of the two has lead some to read chaos theory as a story of order inevitably emerging from chaos (Brion 1995, 181), or of chaos triumphantly defeating order (Hawkins 1995, x), or of their endless cycling back and forth (Demastes 1998; McCarthy 2006, 193). Neither do deterministic systems change their rules as they go (contra Wise 1995, 754; Hawkins 1995, 35). But a temporal dimension is sometimes misread into diagrams that simply show how a chaotic system behaves as a parameter is varied. We are so accustomed to reading the x-axis on graphs as an index of increasing time that we sometimes impose a narrative structure on the depiction of an atemporal relationship.

Enthusiasm for chaos theory sometimes leads people to overstate the prevalence of chaotic behavior. Although research has indeed found strange attractors in a wide variety of natural systems, the fact remains that not everything is chaotic (contra Brion 1995, 184). As Kenneth Knoespel points out, chaos theory is not a metaphysical "prophecy of universal disorder" (1991, 104). Linear systems, for instance, cannot exhibit chaos, but that hardly means that every nonlinear system does (contra van Staveren 1999, 144; and Conte 2002, 21). Neither should we consider any system with feedback (contra Palumbo 2002, 91) or with more than two elements (contra Wise 1995, 754; and Adams, Brumwell, and Glazier 1998, 541) to be necessarily chaotic.

As described earlier, two of the most important features of chaotic systems are sensitive dependence and strange attractors. Each of these concepts has been subject to numerous misunderstandings. One of the most common mistakes about sensitive dependence on initial conditions confuses it with just any sort of sensitivity to small influences. I call this kind of fallacious reasoning the Jurassic Park effect: the fact that one small event can result in disastrous and disproportionate consequences does not mean that *every* small event will yield unpredictable results. The butterfly effect should not be conceived of as a story about the special causal powers of one special butterfly; in a chaotic system, each and every butterfly matters, and that is what produces intractable unpredictability (contra Palumbo 2002, 85; and Slethaug 2000, 62). One of the most common misconceptions about strange attractors is that they are somehow magnetic. But the "attraction" happens only in an abstract mathematical space used to represent the behavior of the system (contra van Staveren 1999, 143; Hawkins 1995, 20; McCarthy 2006, 249). Even if magnetism is used as an illustrative metaphor, the strange attractor is not in any way an outside entity or force but a mathematical depiction of the long-term behavior of the system.[7] Many more errors and confusions will come to light as we explore the ways that researchers have used chaos theory, but the purpose of this book is not to catalog mistakes or to scold people for misunderstanding science. Some errors have consequences for the arguments that engage in borrowing, and it is easier to evaluate those arguments if we have a clearer sense of what is being borrowed.

Mainstreaming and Acceptance

Chaos theory has enjoyed tremendous popularity both in popular culture and in the academic world. By 1989, science writer Ian Stewart could declare that "the striking computer graphics of chaos have resonated with the global consciousness; the walls of the planet are papered with the famous Mandelbrot sets" (1989, 42). Chaos theory has shown up in comic strips, television shows, computer games, and even in a description of Bill Gates's business philosophy (Pitta 1998). By 2004, this once cutting-edge science had been so thoroughly absorbed that a mediocre Hollywood movie was titled *The Butterfly Effect*, and a reviewer said that the title referred to an "old canard" (Kehr 2004). But has chaos theory made any real impact besides giving us fractal coffee mugs and providing an excuse for Jeff Goldblum to flirt with Laura Dern in *Jurassic Park*? Some believe it has had "little effect on the way that nonscientists think about the world" (Matheson and Kirchhoff 1997, 43). I do not seek to evaluate the impact of nonlinear dynamics on our culture as a whole, but it is worthwhile to see that chaos theory has played a significant role in mainstream academic research.

Randall Bausor says that compared to physical scientists, economists have greeted nonlinear dynamics with "less enthusiasm, if not open hostility" and that to many of them it remains "peripheral" and "appears as yet another ephemeral fad" (1994, 109, 120). Yet one of the very sources he cites in evidence of this attitude is a 1989 review article by Brock and Malliaris that states, "we confidently predict that new ideas from nonlinear science and chaos theory will continue to stimulate useful new methodology in economics even though the evidence for deterministic data generators in economic nature may be weak" (339), and "it is not an exaggeration to say that activity in this area is feverish at the present time" (341). More recently, we note the existence of a book introducing methods of nonlinear dynamics to students of economics (Shone 1997) and of journals such as *Journal of Economic Dynamics and Control* and *Macroeconomic Dynamics* as indications of how thoroughly established the "dynamical approach" has become in economics. One of the pioneers of chaos in economics, Jess Benhabib, has served as editor of the *Journal of Economic Theory*; a book bearing the title *Mainstream Mathematical Economics in the Twentieth Century* includes a discussion of deterministic chaos (Nicola 2000); and a special issue of the *Journal of Macroeconomics* was recently devoted to nonlinear dynamics (March, 2006). How mainstream is that?

In the field of legal scholarship we encounter a similarly curious ambivalence. Some have claimed that the law has responded to chaos theory with a "so what?" (Hayes 1992, 751) and that the concept has been unpopular in jurisprudence (Levit 1996, 965). Yet chaos theory has also been said to be "emerging as one of the most-discussed new paradigms in contemporary American thought, both legal and otherwise" (Cunningham 1994, 547).[8] More recently, Trig Smith described chaos theory's "intellectual appeal to the legal community" as actually being a "disadvantage" because it has eclipsed other explanatory possibilities. He goes on to state that "chaos theory appears to be a favored topic in the legal scholarly literature" (2000, 785). And the widespread role of chaos theory in academic scholarship is hardly different in literary studies. Indeed, several of the monographs I discuss appear under their own Library of Congress subject heading, "chaotic behavior in systems in literature." In economics, law, and literature, I cite numerous works by tenured professors publishing in well-respected venues.[9] All of this provides at least some sociological evidence that borrowings from chaos theory cannot simply be dismissed as a fringe phenomenon outside the academic mainstream.

Methodological Challenges, but No Revolution

The rise of chaos theory, and the widespread acceptance of the possibility of unstable and unpredictable behavior in simple deterministic systems, certainly gave rise to some dramatic moments. Physicist J. Lighthill went so far as to say: "I have to speak on behalf of the broad global fraternity of practitioners of mechanics. We collectively wish to apologize for having misled the general educated public by spreading ideas about the determinism of systems satisfying Newton's laws of motion that, after 1960, were to be proved incorrect" (Lighthill 1986, as quoted in Prigogine 1993, 4). More recently, chaos has been heralded as "unquestionably one of the most significant intellectual innovations of the century" (Mirowski 1990, 289) and a discovery that "presents a fundamentally different idea of how to understand reality" (Hayes 1992, 756; see also Brock and Malliaris 1989, 300).

At least since Thomas Kuhn, dramatic upheavals in science often go by the name "scientific revolutions," and there has been no shortage of people describing chaos theory as a revolutionary new paradigm.[10] But, as I have argued elsewhere, chaos theory does not really qualify as a "revolution" or "paradigm shift" in the sense associated with the work of Thomas Kuhn (Kellert 1992; see also Gross and Levitt 1994, 94; Matheson and Kirchhoff 1997, 31; Leiber 1998, 367). It challenges no fundamental physical laws, and the techniques of dynamical analysis have been thoroughly absorbed into ongoing work in physics, chemistry, and biology. Yet it does represent a significant change in scientific methodology, with real challenges to some conceptions about how nature works and about how science ought to go about seeking understanding. Providing an important new tool can present a profound challenge to how we go about our work, but it does not require us to throw away the old tools (Day and Chen 1993, 320; see also Rosser 1991, 324).

In my previous book on chaos theory, I discussed some of these challenges that chaos theory poses for earlier conceptions of scientific practice and scientific understanding (Kellert 1993). The presence of sensitive dependence on initial conditions calls into question the habit of ignoring small disturbances on a system, and chaotic dynamics impel us to reexamine the extent of determinism at work in our world. Furthermore, the study of chaotic dynamics proves a poor match for traditional conceptions of scientific understanding that focus on detailed predictions and causal mechanisms. Chaos theory exemplifies a search for understanding in the form of geometric patterns and long-term behavior. Peter Smith (1998) has suggested that the study of chaotic dynamics does not represent anything especially new in the practice of physics. Although a number of physicists clearly disagree with this evaluation, the historical point is not at issue here. Smith admits that chaos theory poses an important challenge to conceptions of the methodology of physics widespread among nonscientists, including philosophers of science.

What Borrowed Knowledge Is

The term "borrowing" may seem to imply that the natural sciences "own" chaos theory and that scientific knowledge resides in a distinct disciplinary location from which it is transported to another place and (hopefully) returned. But of course no one owns a concept, a theory, or any piece of published knowledge, and the figurative walls or "silos" that divide disciplines do not actually isolate fields of inquiry from one another. I return to the question of disciplines and their boundaries in the next chapter. For now, it is worth noting that the phenomenon of borrowing has been going on for quite a while. Economist Herbert Simon pointed out in 1959 that "the social sciences have been accustomed to look for models in the most spectacular of the natural sciences," and he went on to say "there is no harm in that, provided that it is not done in a spirit of slavish imitation" (as quoted in Prigogine 1993, 4). Knowledge has always traveled between disciplines, from Darwin's use of geological facts to the role of radioactive dating in archaeology to contemporary discussions of historical "forces" and social "inertia" (see, e.g., Klein 1990, 85). The insights of linear Newtonian physics have sometimes proven useful for conceptualizing human social change. If these uses are legitimate, then reconceptualizations of the physical world may lead us to rethink our picture of the social world.

In her introduction to an early and influential collection of essays about chaos theory and literature, Katherine Hayles posed several questions about borrowed knowledge that this study addresses. Because chaos theory challenges traditional presuppositions of science, she asserts that it has significant implications for the broader culture. But how exactly do challenges to science yield consequences for humanities? And who draws these implications? As Hayles points out, "where to draw the line between legitimate and illegitimate extrapolations from chaos theory is not easily or lightly answered" (1991, 15). In this book, I examine how that line can be drawn. Hayles goes on to ask how we are to distinguish appropriate uses of terms such as "nonlinear" that have value connotations within as well as outside of the sciences. She declares that she will not argue for an answer: "rather, I want to urge that the question cannot be dismissed out of hand" (1991, 18). In this work I seek to develop ways to answer just this kind of question. I do not, however, deal with the rhetorical or conceptual influence of literature, economics, or law upon the natural sciences, although I recognize that borrowing goes in both directions. The Austrian economist Friedrich Hayek, for example, identifies a number of concepts relating to human phenomena that the natural sciences were "forced to use" and attempted to free from "anthropomorphic connotations": law, cause, function, order, organism, and organization (1952, 208). In describing the analogies of structure that mediate cultural influences upon scientific models and theories, philosopher of science Helen Longino has spoken of the need to trace "the seepage of language and meaning from one domain to another" and to study "the uses to which the models are put" (1993, 115). In this book, I undertake just such a task but in the other direction: from the natural sciences outward.

The Dangers of Borrowed Knowledge

Physicist David Ruelle alerts us to the pitfalls of borrowing when he writes that "to pursue the discussion at a purely metaphysical and literary level is like driving a car blindfolded; it can only lead to disaster" (1991, 120). Yet if discussions of the natural sciences at the "literary level" cannot and should not be ruled out of bounds, how can we identify the obstacles that lie on this road? Law professor Royce de R. Barondes has expressed concerns about the perils of borrowing knowledge from chaos theory for legal scholarship: "hazards exist in integrating theories of one discipline into another. The elegance of a contained, well-defined theory may suggest more grandiose applications of questionable validity" (1995, 161). Barondes identifies five potential dangers in the use of scientific terminology: (1) the "invitation of scorn for imbedding meaningless comparisons in the language of legal scholars"; (2) "application of scientific labels may substitute for a considered analysis"; (3) "misleading labels frame the inquiry in a manner that biases the conclusions and deceives scholars into disregarding certain lines of thought"; (4) "such terminology facilitates lexical legerdemain, obfuscating logical leaps"; and (5), the practice "engenders in the reader a false sense of understanding the scientific area" (177). In what follows, we will have occasion to notice these problems arising in a number of cases. But we will also see examples of borrowing that escape these risks.

The physicists Alan Sokal and Jan Bricmont also raise warnings about borrowed knowledge in their 1998 book *Fashionable Nonsense: Postmodern Intellectuals' Abuse of Science*. They claim that some famous French intellectuals have "repeatedly abused scientific concepts and terminology," either using them out of context without justification or throwing around jargon without concern for its meaning or relevance (Sokal and Bricmont 1998, x). But they do not condemn borrowed knowledge out of hand: "we are not against extrapolating concepts from one field to another, but only against extrapolations made without argument" (x). I would go even further: I am not only concerned about extrapolation made *without* argument; I am concerned about extrapolations made with *bad* arguments. It is worthwhile and important to critically examine such arguments when they are given and to seek to reconstruct them when they are not. Thus, my project takes up where Sokal and Bricmont's leaves off. They state that they are after "name-dropping, not just faulty reasoning. Thus, while it is very important to evaluate critically the uses of mathematics in the social sciences and the philosophical or speculative assertions made by natural scientists, these projects are different from—and considerably more subtle than—our own" (Sokal and Bricmont 1998, 15). Subtly or not, I have in the past examined some physicists' speculations about chaos theory (Kellert 1993), and in this work I take up the other project mentioned by Sokal and Bricmont. Joe Moran agrees that the use of science by French postmodernists has not been the most helpful form of interdisciplinary endeavor. He recommends that a better project "might be constructed by examining the ways in which scientific ideas extend beyond the area of specialized inquiry and form part of culture . . . how they interact with 'non-science' when they are widely disseminated" (2002, 159). I seek to build a more productive interdisciplinary space by doing just that.

Beyond the hazards of name-dropping and erroneous terminology lies an even more serious issue involved in the borrowing of knowledge from the natural sciences: the "spirit of slavish imitation," as stated by Herbert Simon and quoted earlier. The unwholesome influence that the natural sciences sometimes exert as knowledge travels back and forth across disciplinary boundaries was well characterized by Hayek. He wrote that the success of the natural sciences was so great that

> they soon came to exercise an extraordinary fascination on those working in other fields, who rapidly began to imitate their teaching and vocabulary. Thus the tyranny commenced which the methods and technique of the Sciences in the narrow sense of the term have ever since exercised over the other subjects. These became increasingly concerned to vindicate their equal status by showing that their methods were the same as those of their brilliantly successful sisters rather than by adapting their methods more and more to their own particular problems. (1952, 13–14)

Hayek uses the term "scientism" to name this "slavish imitation of the method and language of Science" (1952, 15). In what follows, I am more concerned with the imitation of the language, concepts, and results of chaos theory as opposed to its methods. Outside of economics, the quantitative methods of chaos theory have found little application in the humanities and social sciences (although there have been attempts to subject legal and even literary fields to quantitative analysis, usually with unfortunate results).[11] But whereas my target is not the aping of a supposed "scientific method" (the concern of, for instance, Priest [1981], in law), we will see a number of instances in which chaos theory has been used to argue for a shift in the methodology of other fields. Hayek identifies a phenomenon of cross-disciplinary enthusiasm that leads people to "overstrain a new principle of explanation. . . . Gravitation and evolution, relativity and psycho-analysis, all have for certain periods been strained far beyond their capacity" (1952, 208). I am interested in asking why researchers do this, and how we can tell when an idea's "capacity" has been exceeded.

The Limitation to Three Fields

It would be almost impossible to survey all the uses of chaos theory outside the natural sciences. My more narrow focus—on English-language works in economics, law, and literature—nonetheless allows us to look at technical applications and metaphorical speculations, as well as at applications to evaluative issues of public policy and art. Economics is often seen as among the "hardest" of the social sciences, and literary studies maybe ranked among the "softest" of the humanities, with law somewhere in between. Although I do not support this familiar ordering of disciplines based on their metaphorical hardness, it is important to recognize that this ranking plays a role in how disciplinary prestige is accorded and how borrowing is viewed. For example, Frank Miele, the senior editor of *Skeptic* magazine, introduced a special section on chaos and complexity theory by bemoaning a "lack of accountability" in uses of chaos theory that increases "as one moves from the physical sciences, to the biological, to the behavioral, to the social sciences, reaching the level of a dead cert in literary studies" (2000, 54). Indeed, sometimes the term "science" is used to refer only to the natural sciences. I must apologize in advance for occasionally slipping into this pattern of usage that excludes the social sciences.

In an atmosphere in which disciplines receive admiration or suspicion based on how close they are to physics, some may wonder whether nonscientific disciplines such as law and literature produce any genuine knowledge at all. Whereas one literary scholar has suggested that "scientific knowledge can be used to construct knowledge of the arts" (Slethaug 2000, xiv), others may contend that fields such as literary study are merely academic subcultures that pretend to the status achieved by the natural sciences. Indeed, the humanities have sometimes been treated as nothing more than holding pens for unsystematic considerations of phenomena that do not yet yield to our current experimental methods. In this work I do not adjudicate invidious disputes between disciplines because the work of legal and literary scholars still matters, even if it is denied the honorific title "knowledge." Research in the social sciences and humanities bears on questions about how to arrange human affairs and articulate human experience, and this makes it important to attend to the actual arguments made in those fields. In the remainder of this introductory chapter, I briefly discuss the contexts within which chaos theory has been borrowed in each of these three fields. Of course, these disciplines interact with each other as well as with the natural sciences. Subfields known as "law and economics" and "law and literature" are well established, and although I do not know of a "literature and economics" area, the "rhetoric of economics" movement comes close.[12]

Economics

In an early review article, economist David Kelsey observes that "economics and weather forecasting have a lot in common. When people are not talking about the weather they talk about the economy. Both weather forecasters and economic forecasters have a bad name with the public" (1988, 1). But he notes some further similarities that have made economics ripe for the application of techniques from nonlinear dynamics: both are complicated systems with apparently random behavior that nonetheless displays patterns. Economists have borrowed from chaos theory in three main areas: theoretical models that show the possibility of chaotic behavior in economic systems, quantitative techniques for examining whether actual economic data reveal chaotic behavior, and examinations of the implications that chaos theory has for the way economics is practiced and the way economic policy ought to be pursued. Sometimes the economic systems under discussion are microeconomic (such as the markets for hogs, Microsoft stock, or pounds sterling), whereas other times they are macroeconomic (such as national rates of unemployment, inflation, and investment).

Much of what economists do consists of constructing abstract mathematical models of economic systems and then proving theorems about the behavior of these models. And some of the first applications of chaos theory to the social sciences came when economists demonstrated that economic models could exhibit behavior that satisfied the mathematical definition of chaos. In macroeconomics, much work has focused on models of the "business cycle," a long-standing puzzle about what is responsible for large fluctuations in macroeconomic measures such as the Gross Domestic Product. Conventional economic thinking assigns these fluctuations to the random alignment of myriad individual economic agents, or to extrinsic political "shocks" (such as regulations or embargoes) delivered in distressingly stochastic fashion. Chaos theory supplies an alternative account: perhaps the fluctuations arise from the intrinsically unstable dynamics of a deterministic system with a relatively simple (or "low-dimensional") chaotic attractor.

The investigation of nonlinear dynamics in economics raises important questions about how to do economics and how to create governmental policies. Do the theoretical models suggest that our general approach to economic modeling ought to be changed? Does the possibility of chaos carry policy implications? These methodological and prescriptive questions constitute a main focus of my discussion about chaos theory and economics in the chapters to follow. But there are other, more technical issues. For instance, are the parameter values required to generate chaos in the mathematical models so outlandish that such behavior represents a mere possibility, never to be expected? This question lies beyond my expertise. Do we find any empirical evidence that actual economic systems display chaotic behavior? This question of the real-world applicability of chaos theory has given rise to an active field of research seeking to discern signs of chaos in economic data. The refinement of statistical techniques for determining the presence of chaos has been the site of lively disagreements over very technical points that I am not capable of judging. Some may hold that no one should dare to comment on issues of economic methodology or policy unless they are fully competent to do publishable research on the related technical issues. But surely that sets the bar too high for participation in discussions that affect us all. In fact, a recurring topic in my examination of borrowed knowledge is the question of who ought to be allowed to engage in discussions about the implications of science.

Law

Many attempts to use chaos theory in legal scholarship address the evolution of the common law tradition. The body of doctrine in common law does not grow by passing new legislation; instead, it evolves according to the decisions made by judges who seek to follow precedent but sometimes introduce new guidelines or even radical changes. As we will see, several scholars have borrowed concepts from chaos theory as a way to understand the dynamics of this process, and this use has far-reaching implications for how law is practiced, how it is taught, and how it is studied.[13]

Yet the natural sciences influenced the law long before chaos theory came on the scene. Indeed, the United States Constitution was designed to be analogous to a Newtonian machine with forces and counterbalancing forces (Tribe 1989, 3). But it was not until the twentieth century that some have attempted to make the law itself "scientific." These efforts began with Christopher Langdell and his endeavors to professionalize American legal education at a time when many believed that a field of study had to be a science in order to be worthwhile (Williams 1987, 430). Imitating the method that science supposedly follows, Langdell thought that by analyzing legal phenomena, "he could discern the law's underlying principles, develop doctrinal rules that would reflect those principles, and predict future decisions" (Veilleux 1987, 1975).

In the 1981 symposium "Interdisciplinary Study of Legal Evolution," Robert Clark declared that such research "should be more scientific in aspiration than are most historians' accounts of legal change: it should seek to identify and to test lawlike generalizations about changes in specific fields of law" (1981, 1238).[14] But he too warns of the possible dangers of borrowed knowledge: "the legal scholar that tries to go only part of the way, borrowing only the bare theoretical concepts and explanatory schemes of another discipline (for example, economics) and routinely applying them to legal phenomena runs the risk of trying to pull rabbits out of an empty hat, and of deluding himself into thinking that he has done so" (1261). Of course, pointing out that an activity runs a risk does not yet count as criticism.

These quotations from Clark alert us to one of the areas in which cross-disciplinary work connects two of the fields I examine: law and economics. Indeed, a movement by the name "law and economics" sees the law as influencing human behavior according to predictable patterns and explicitly uses economic concepts such as "efficiency" in order to study the law (Scott 1993, 343; Downs 1995). As might be expected, the discipline of economics has greatly influenced the law and public policy with regard to such issues as antitrust regulation (Wise 1995, 714). Another connection between fields is provided by the "law and literature" approach, which can be traced back to James Boyd White's 1973 *The Legal Imagination: Studies in the Nature of Legal Thought and Expression*. It will be helpful to remember the words of John Veilleux with regard to these approaches that link the law with science or with literature: their role "is not to be a *source* of greater substantive knowledge of the law; it is instead primarily to be an intellectual *device* to promote and facilitate that understanding" (1987, 2000–2001). This distinction resurfaces during an examination of the different rhetorical functions served by borrowing.[15]

Literature

Connections between chaos theory and literature take many forms. Because my focus is the making of knowledge rather than the making of art, I do not direct my attention to works of literature that make explicit use of chaos theory, such as Tom Stoppard's *Arcadia* or *The Last Voyage of Somebody the Sailor* by John Barth.[16] I pay some attention, however, to the use of chaos theory to read particular works of fiction, although I do not claim any final authority on questions of what counts as a good, illuminating, or convincing reading of a literary text. I concentrate mostly on the uses of chaos within the field of literary theory, which more nearly parallels the borrowing within other academic fields of knowledge production and argumentation. For, as mentioned previously, I focus less on uses of chaos theory to analyze particular court cases or economic data and more on the role of nonlinear dynamics in arguments within jurisprudence and economic theory or policy. It is also worth bearing in mind that, with regard to literature, the term "theory" does not refer to a systematic account that should be judged by predictive success but rather "a set of speculative statements that serve as guides to reading and interpreting texts" (Hayles 1990, 36).

Chaos theory does not represent the first attempt to use science to study literature, but it enjoyed a tremendous vogue during the 1990s. In their review of this work, Matheson and Kirchhoff concluded that "the application of chaos theory to literature has to date produced only a homogeneous and quite unexceptional set of assertions, no matter which texts have been under consideration" (1997, 39). Surely they are correct that there is a lot of mediocre work published on chaos and literature, but there is mediocre work in every field. In their analysis, they address three main claims:

> (*a*) There are significant similarities between the science of chaos, contemporary literary theory, and/or philosophy.
> (*b*) We can trace a common etiology for chaos theory and poststructuralist criticism. The two are so intimately related that they are really different formulations of a common claim, as filtered through two different fields.
> (*c*) The science of chaos can help us interpret and understand specific literary works, and perhaps contemporary literature in general. (1997, 29)

In what follows, I am primarily concerned with claim (*a*), and I also have some comments about claim (*c*) with regard to particular works.

For although Matheson and Kirchhoff admit that it is not impossible for chaos theory to play an illuminating role in literary interpretation, they do not seem to find any actual examples (1997, 43). I have the least to say about claim (*b*), which concerns a broader cultural analysis of chaos theory and its relationship to other fields. Some view the relationship in terms of science and literature responding to a shared cultural climate (Hayles 1990, 184–85; Conte 2002, 143), whereas others consider that the natural sciences influence, direct, or even determine the character of artistic expressions (e.g., Slethaug 2000, x).[17] Still others hold that the influence runs in the other direction, with literature anticipating or even making possible new discoveries in science (see Hayes 1992, 763–64; Demastes 1998, 8). But my goal is to evaluate literary arguments that borrow from chaos theory, not to discern the direction of causal influences in the culture at large.[18]

Overview

Some traditional conceptions of knowledge make the phenomenon of borrowing seem epistemologically uninteresting or essentially illegitimate. Physicist Steven Weinberg illustrates such a view when he asserts that "those who seek extrascientific messages in what they think they understand about modern physics are digging dry wells," because the knowledge we get from physics "has no legitimate implications whatever for culture or politics or philosophy" (1996, 12).[19] My goal is not to advance technical arguments against the philosophical positions that animate such a view but to outline and defend an alternative conception that makes possible important critical work. So I take a twofold approach in the chapters to come. First, I challenge one or more of these traditional conceptions in order to create a space in which the critical examination of borrowing can take place. Then I demonstrate the possibility of such critique by actually engaging in the substantive evaluation of particular instances of borrowing from chaos theory. If some of these specific criticisms fall short, this flaw should not invalidate the more general points.

Among the widely accepted conceptions I seek to challenge is the notion that the process of generating ideas should be considered entirely separate from the process of testing them. This view would relegate borrowed knowledge to being merely an inspirational tactic, unworthy of serious critical attention. Similarly, if fields that describe how knowledge is made cannot help us answer prescriptive questions about what should count as knowledge, this inability would severely restrict the kinds of approaches considered appropriate for investigating borrowed knowledge. In chapter 2, "Disciplinary Pluralism," I address these misguided conceptions while considering the question of what approach should be used to understand the phenomenon of borrowed knowledge. Indeed, what makes us think we should use only one approach to understand this, or any other, phenomenon? I defend the use of a variety of approaches from multiple disciplines, responding to concerns about whether interdisciplinarity is unsound and whether empirical studies can help us in the task of evaluating how well borrowed knowledge works.

Another widely accepted conception that would limit our ability to examine cross-disciplinary work is the notion that social and persuasive aspects of the knowledge-making process are unfortunate features that serve only as sources of possible contamination. Researchers often borrow knowledge from the natural sciences to help them convince others to take them seriously. But if "persuasion" is thought of only as emotional manipulation, then an examination of the persuasive functions served by borrowing can tell us only about academic marketing strategies and nothing about knowledge itself. In chapter 3, "The Rhetorical Functions of Borrowing and the Uses of Disciplinary Prestige," I develop and defend the use of a rhetorical approach to examine some of the ways that cross-disciplinary research does its persuasive work. Borrowing can help researchers have their efforts read and taken seriously, in part because of the prestige of the natural sciences. I explore some of the ways that authors harness this prestige and some of the dangers posed by its use. But although some of the rhetorical functions of borrowing may be unsavory, it can also serve a useful purpose, as I show in chapter 4, "Motivating Methodological Change." In that chapter, I examine several cases in which researchers have enrolled the prestige of the natural sciences to help them motivate substantial methodological changes in their home discipline.

Some have dismissed all uses of chaos theory by researchers in the social sciences and humanities as merely metaphorical—at best an incomprehensible source of inspiration and at worst a source of nothing but confusion. But if metaphors are denied even the possibility of conceptual function, then in many cases the crucial task of critiquing borrowed knowledge will not be able to get off the ground. So my next two chapters investigate the metaphorical uses of chaos theory in light of recent understandings of how metaphors work. In chapter 5, "Metaphorical Chaos," I explore the role of metaphor in making knowledge and defend the metaphorical use of chaos theory as sometimes useful. Chapter 6, "How to Criticize a Metaphor," considers in detail some of these

metaphorical uses. I develop a new framework for the critical evaluation of metaphors and illustrate its application to a number of examples.
The applicability of this framework demonstrates that we can, in fact,
subject metaphorical borrowing to critical scrutiny.

One final conception I resist is the rigid dichotomy between facts and
values that insists science can have nothing to say about ethical, political,
or aesthetic questions. Such a view would make all attempts to use chaos
theory to address such questions simply fallacious and unworthy of further analysis. Chapters 7 and 8 explore whether knowledge borrowed
from the natural sciences helps us answer questions about the aesthetic
worth of a work of literature or the rightness of an economic regulation.
In chapter 7, "Facts, Values, and Intervention," I argue that questions of
value should not be simply reduced to scientific questions of fact, but that
scientific knowledge can nonetheless help us answer questions about better and worse social arrangements. The focus here is on questions of governmental intervention in economic and social affairs. And in chapter 8,
"Beautiful Chaos?" I explore some attempts to use chaos theory as a
basis for aesthetic evaluation.

Chapter 9, "Postmodern Chaos and the Challenge of Pluralism,"
explores the purported connections between two realms of inquiry that
enjoyed great popularity in the 1990s: chaos theory in the sciences and
postmodernism in cultural studies. The numerous attempts to connect
these two areas provide an opportunity to recapitulate and synthesize the
issues about disciplines, rhetoric, metaphor, and values that have come
before. I conclude by revisiting the main lessons learned about how disciplines can and should interact.

Throughout this book, I seek to defend a pluralistic and roughly
pragmatist conception of knowledge by showing that it enables and
advances the critical examination of borrowed knowledge. As a preliminary characterization, allow me to present my pluralism in somewhat
unconventional terms.[20] As we make knowledge, we encounter many
things that may seem odd, transgressive, or even threatening: transfers
across disciplinary boundaries, nonstandard usages of words, traffic between facts and values, and so on. Pluralism counsels that we should
not limit our options in such encounters to either fearful rejection or
uncritical embrace. The first option may lead to a rigid reassertion of
boundaries that enforce isolation, whereas the second may result in the
inappropriate collapsing of those very boundaries. New possibilities do
not have to be pigeonholed as either menacing or liberating but should
be critically examined for what they offer and where they fail. Pluralism
recommends a number of virtues for the conduct of inquiry. Respect and

tolerance should be shown by recognizing that disciplinary approaches incompatible with your own may serve an important epistemic purpose, and humility is shown by not assuming that your preferred approach or theory is sufficient. Responsibility is displayed by making oneself answerable to other inquirers and to the subject matter inquired into, as well as by taking other disciplines seriously enough to learn what they actually say. And open-mindedness is shown by a willingness to evaluate proposed methods or results even if they conflict with your preferred view. A pluralist approach can help to articulate and illustrate these intellectual virtues, and in this way the examination of borrowed knowledge can serve as a case study that shows the need to pursue our inquiries in a more balanced and broad-minded spirit.

2 Disciplinary Pluralism

Many disciplines provide useful ways to understand and evaluate the phenomenon of borrowing. By turns, we may have to ask questions about the social structures of academic work, the way language works, and the nature of evaluative questions. Before addressing these specific tasks, however, I articulate and defend my general methodological approach. This methodology has two main features: *disciplinary pluralism*, that is, using a wide array of techniques of inquiry, and *normative naturalism*, that is, using empirical inquiries to inform evaluative judgments about what works well and what does not. After clarifying what I mean by methodology in this study, I turn to disciplinary pluralism and then to normative naturalism. Along the way I do not break new philosophical ground so much as articulate and defend a particular mode of interdisciplinary endeavor that seeks neither to unify all disciplines nor to keep them insulated from each other. As reader, you may already accept that borrowing is a complicated phenomenon requiring several different approaches in order to understand how it happens, and why. You may also already accept that using these various approaches will help us make judgments about when borrowing works well and when it works badly. If so, consider this chapter both a general argument for a pluralistic approach to interdisciplinary work and a sketch of positions to be argued in later discussions.

Methodology and Metamethodology

Methodology is the examination of a research process that asks how that process works as well as how it should work. A methodology gives rules, techniques, guidelines, and suggestions—not at the level of which stain to use to see a particular organelle in a plant cell or how to translate a particular term from the Latin, but at the level of a broad, general, and theoretical investigation of the conduct of inquiry. So methodology includes both descriptive and evaluative elements, involving questions of fact about the knowledge-creation process as well as judgments about improving that process and appraising its results. Epistemology, the philosophical theory of knowledge, also deals with questions about knowledge from a theoretical perspective, although epistemology is typically even more general and abstract than methodology. The line between methodological concerns and epistemological ones need not be a sharp one, however.

The methodology used in this book is thus an overall theoretical framework for examining borrowing. Confusion may arise because my object of study, borrowing, is itself a phenomenon that takes place within others' research process. Thus, we are undertaking a methodological inquiry when we ask, "How does borrowing take place?" and "When does it work well?" But I am not asking those questions in this chapter; they are the primary subject of the chapters to come. Before asking these questions about borrowing, I look at how to go about answering them. A brief explanation of the different levels of discourse involved in the questions will help keep the distinction clear and help us avoid piling up too many "meta's."

At level 0, picture the things that are the objects of study—atoms, for instance, or commodities markets. I make no assumptions about the ultimate metaphysical status of these entities; for the purpose of this classification scheme, it makes no difference that some entities at level 0 may be independent of human activity whereas others may be entirely "socially constructed." And the designation of this level as 0 should not suggest that this is the most (or least) "important" level; no ontological primacy or epistemic priority is intended. At level 1 we have disciplines that investigate objects of study—physics studies atoms, for instance, and economics studies markets. These disciplines ask first-order questions— questions about things one level lower on the scale. So physics asks: How do atoms work? How can we get them to do what we want them to do? Do they behave deterministically or indeterministically? And so on. Economics asks: How do markets work? How can we get them to do what we want them to do? Should they be free or regulated? And so on. Economists may look over at the physicists doing their first-order work, borrow some of their concepts or tools, and then use them to look back at their own objects of study in a new way. This is just the phenomenon of borrowed knowledge; note that it occurs here on level 1.

Level 2 is where methodology (and epistemology) live. Here, people ask questions about the disciplines one level lower. Thus, the methodology and philosophy of physics ask questions about physics: How does physics work? How should it go about its business? Does it give us true representations or just useful fictions? Another characterization of this level might see it as asking second-order questions about the things two levels beneath it: How should we go about answering questions about atoms? Similarly, the methodology of economics lives on level 2, asking such questions as: What is the role of rhetoric in economic research, and what standards should we use to make judgments about whether markets are working efficiently? Questions about the relationship between level-1 enterprises get asked at level 2: How do physics and economics interact? Because cross-disciplinary borrowing, the subject of this book, is a phenomenon at level 1, most of the questions I address in later chapters are level-2 questions: How does borrowing work? And, when does it work well?[1]

Level 3 is the realm of metamethodology and metaphilosophy. Here we encounter such questions as How should issues in the philosophy of science be resolved? Should we appeal to a priori intuitions or use some other technique? And, what should be the relationship between different level-2 inquiries? How should the sociology of science and the philosophy of science resolve their sometimes-conflicting accounts of scientists' behavior? In this chapter I propose one particular approach to investigating the phenomenon of borrowing. The present chapter therefore asks second-order questions about a level-1 subject matter, so these considerations belong to level 3: How shall we go about understanding borrowed knowledge?

To those who grow weary at the thought of meta-meta-level discussions, I can only apologize. But the distinction between levels is important, for although in this chapter I deal with the relationship between different disciplines, I am dealing with this relationship at level 3 only. That is, I am not asking a level-2 question such as What should be the relationship between biological explanations of human behavior and

ethical explanations of human behavior? Instead I discuss the relationship between disciplines as they investigate other disciplines themselves;
I engage in level-3 questions such as What should be the relationship
between *sociological* investigations of cross-disciplinary work and *epistemological* investigations of cross-disciplinary work? Therefore, I do
not deal directly here with questions about the relationship between science and society, although what I have to say in this chapter is relevant
to such questions. In fact, this chapter foreshadows on the meta-level a
view about the relations between science and society, which I defend in
later discussions. The first level-3 question I address is the following: In
seeking to understand borrowed knowledge, should we insist on using
only one discipline? As an advocate of disciplinary pluralism, my answer
is no.

Disciplinary Pluralism

My argument for disciplinary pluralism draws much inspiration from
pluralistic positions within the natural sciences themselves, which reject
the assumption that comprehensive accounts are always possible and
desirable. Explanations of the evolution of sexual selection, or of human
behavior, or of quantum-mechanical dynamics may best be pursued using
multiple approaches that are actually incompatible with each other (see
the essays in Kellert, Longino, and Waters 2006). The pluralist stance in
science recognizes that different theoretical structures and experimental
approaches work well to explore different aspects of a scientific phenomena, and that we should not presume that all these structures and
approaches can or should be integrated into one comprehensive scheme.
In a similar vein, Lawrence Cunningham argues that market phenomena
are so complicated that they are probably not best approached with a
single economic account (1994, 607), and William Paulson urges that
a literary text cannot be adequately described at one level, or by one
discipline—not linguistics or rhetoric or narratology or psychology or
politics (1991, 48).
    The scientific pluralist stance defends the peaceful coexistence of in-
compatible theories and approaches within a discipline. Similarly, disciplinary pluralism invites the use of techniques from multiple disciplines
to understand the subject matter of an investigation. These different
techniques may be used by one investigator, by a team of researchers, or
by a number of individuals or teams. The "division of cognitive labor"
(see Kitcher 1990) often provides a valuable research strategy, so particular investigators or teams may each pursue their own disciplinary

approach. These approaches may combine, cooperate, compete, or remain aloof from one another—a genuine pluralism does not require any particular form of interaction among disciplines. In fact, a fully pluralistic approach admits the possibility that for some questions the number of disciplinary approaches required is exactly equal to one. To borrow a technical term, we might label a non-imperialist monodisciplinary approach, that remains open to the possibility of other disciplines being relevant, as a "degenerate case" of disciplinary pluralism.[2]

What Is a Discipline?

In this discussion, I use a relatively broad sense of what counts as a discipline. Following William Bechtel's overview, disciplines can be identified and distinguished by their objects of study (domains, phenomena), by their cognitive tools (theories, techniques), or by their social structure (turf, journals) (Bechtel 1987, 297; see also Hayles 1990, 191). Using this broad framework, we can see that other characterizations of disciplines focus on one or more of these three criteria. Lindley Darden and Nancy Maull (1977, 44) and Joseph Kockelmans (1979, 127) consider objects of study and cognitive tools but exclude considerations of social structure. Moti Nissani, on the other hand, defines a discipline almost entirely in terms of social structure as "any comparatively self-contained and isolated domain of human experience which possesses its own community of experts" (1997, 203). Although Stephen Toulmin includes all three features, he reserves the term "discipline" for those systematic endeavors with a clear agreement on central problems and ways to solve them, characterizing atomic physics and law as disciplines but excluding philosophy (1972, 145). In this volume I talk about disciplines in a broader sense in which they are knowledge-producing enterprises with some shared problems, with some overlapping cognitive tools, and with some shared social structure. Thus, I call philosophy and sociology "disciplines," although there is certainly no set of shared techniques that all practitioners of these fields share. Consider a "Continental" philosopher such as Edward Casey (1997) and an analytic philosopher of science such as John Earman (1989) discussing space, for example. They have little in common in terms of style of argumentation or overlapping references.

In some sense, then, pluralism needs no defense and requires no one to advocate for it. In fields such as science studies, as in area studies or women's studies, multidisciplinarity is already firmly emplaced, and interdisciplinary work "has almost become *de rigueur* in contemporary legal scholarship" (Moran 1997, 155). And, of course, chaos theory itself represents a profligately multidisciplinary enterprise, with its roots in mathematics, meteorology, and population ecology. The current state of knowledge production even in the most traditional disciplines is rife with a diversity of methods and marked by a thoroughgoing opportunism. Indeed, Ira Livingston contends that some disciplines such as literary study seem to break down when examined closely, and in fact depend for their existence upon switching and traffic across their boundaries. Using an image borrowed from chaos theory, Livingston describes disciplinary boundaries as "fractal interzones" whose fuzziness "makes them peculiarly recalcitrant as much as it makes them peculiarly malleable" (1997, xi; see also Abbott 2001). But whereas disciplines may be fluid, multiple, or even fractal in practice, such a condition is far from their professed ideology. All too often we operate with a picture of unbridgeable differences in objects of study or technique, and we conceive of interdisciplinary interaction as the cooperation of these essentially isolated endeavors. My defense of disciplinary pluralism therefore needs to focus on normative questions about the relationship between disciplinary methods of inquiry: Given the fact of the plurality of disciplines, what ought we do? Shall we unify them, or isolate them, or bring them into fruitful interrelationships? Pluralists will hold that interactions between disciplines sometimes call for intimate relationships, sometimes for marriages of convenience, and (to extend Ronald Giere's [1973] metaphor) sometimes for a divorce.

Although disciplines are knowledge-producing enterprises, they are much more. Academic disciplines are also social structures that produce, distribute, and control (even "discipline") a great number of things: students, prestige, power, money, access to public debate, and large, large quantities of texts. Interdisciplinarity carries the risks and the rewards of challenging these existing modes of organization. Indeed, it sometimes seems that university administrators show a special interest in interdisciplinary endeavors precisely because they subvert the considerable institutional autonomy vested in academic departments (see also Moran 2002, 183). However, disciplinary pluralism along the dimension of social structure is not the focus of this discussion. I do not seek to defend the multiplicity of academic departments, journals, or degree-granting programs, for instance; neither do I seek to defend a pluralism along the dimension of domains of study. This ontological question of whether different fields have irreducibly different objects of inquiry must be left aside. Of course, we may have metaphysical inclinations that guide us to consider some disciplines inappropriate for some subject matters. Many consider theological inquiry to be unnecessary in botany, whereas others confidently assert that number theory calls for exactly one disciplinary approach. But the question of which discipline or disciplines are appropriate ought not be settled by prior metaphysical intuitions about the intrinsic nature of the subject matter. That question is an empirical, contingent matter, in which our ontological convictions play the role of starting points rather than definitive conclusions.[3]

Disciplinary Pluralism as a Variety of Interdisciplinarity

In articulating and defending pluralism with regard to disciplinary approaches, it is helpful to clarify the distinctions between interdisciplinary efforts narrowly conceived and the broad range of activities that are called interdisciplinary. A variety of terminology finds use here, but most of it aligns in a way that helps us make a few useful distinctions. First, there are *narrow-sense interdisciplinary* efforts that involve creating a new discipline between two existing ones, perhaps leaving the two original disciplines unchanged. In the sciences, Darden and Maull call such new disciplines as bio-physiology "interfield theories" (1977). *Multidisciplinary* work, such as that found in area studies, involves the juxtaposition of two or more disciplines; the different perspectives of the different disciplines are cumulative but not highly interactive, so there is little mutual change, combination, or integration. The term "*cross-disciplinary*" provides the most apt term for the main subject of this book—the borrowing of knowledge from one field in order to assist the endeavors of another discipline. Finally, *transdisciplinary* approaches are more comprehensive, looking for the unity of an overarching synthesis in the grand and sweeping manner of Marxism, systems theory, sociobiology, and so forth (see Klein 1990; Kockelmans 1979).

In her survey of theories of interdisciplinarity, Julie Thompson Klein highlights a contrast between what she calls the "instrumental" and the "synoptic" views of interdisciplinarity. The former view considers interdisciplinary efforts as a practical matter for solving problems and is likely to be satisfied with multidisciplinary approaches, whereas the latter is motivated by a philosophical commitment to coherence and unification and longs for transdisciplinary synthesis (1990, 42). Klein goes so far as to say that "all interdisciplinary activities are rooted in the ideas of unity and synthesis, evoking a common epistemology of convergence" (11). Indeed, many authors who borrow chaos theory display a fondness for vague invocations of a convergence between the sciences and the humanities (e.g., Rice 1997, 106ff.). Literary theorist Alexander Argyros (1991) displays a full-blown transdisciplinary orientation when he seeks to construct a speculative transdisciplinary synthesis of philosophy, art, anthropology, biology, and the physical sciences. As we see in later chapters, Argyros does not limit himself to straightforward cross-disciplinary borrowing that would simply use chaos theory to interpret texts or argue about literary theory. He constructs a grand, comprehensive, and occasionally odd synthesis that then forms the basis for his declarations about the nature of reality, knowledge, society, religion, and art. John A. McCarthy provides a more recent example of transdisciplinary synthesis in his 2006 book *Remapping Reality: Chaos and Complexity in Science and Literature*. In this work, he seeks to elucidate the deep structures that underlie physics, biology, psychology, ethics, and literature. By way of contrast, Jeffrey Rudd has recently criticized the efforts of legal theorist J. B. Ruhl to use chaos theory as the basis for a unification of sociolegal systems with biological and physical ones. Rudd urges us to "wave goodbye to all-encompassing, unifying theories once and for all" because their pursuit threatens to lead us into a dangerous tyranny of experts (2005, 618).

Pluralism about disciplinary approaches does not reject transdisciplinary efforts out of hand but refuses to accept a presumption that unity is always possible or synthesis always useful. A pluralist does not presume that all disciplines are commensurable and ultimately unifiable, but also does not presume that each discipline is necessarily isolated and incommensurable with all others. Genuine pluralism holds that for any two disciplines, it is an open question whether they can or should be united or crossbred. Such questions must be answered case by case, and the answers depend on the usefulness of interdisciplinary efforts for addressing the problem at hand. For some problems, a multidisciplinary approach of cooperation without integration is the best. For other problems, the creation of a new discipline that combines two or more others may give rise to a useful approach. And in still other situations, what is most helpful is the unification of a number of disciplines that are then subsumed in a transdisciplinary synthesis. But pluralists hold that there is no particular reason to presume that isolation, cooperation, crossbreeding, or synthesis is good in all cases.

One way to make a preliminary case for disciplinary pluralism in the examination of borrowed knowledge is to sketch some general advantages to an interdisciplinary approach. As philosopher of science William Bechtel points out, sometimes there is a fruitful interchange between practitioners of disciplines who realize that they have interpreted the same phenomenon in radically different ways (1987, 299). In fact, Moti Nissani has cataloged a long list of advantages of interdisciplinarity, which includes bringing about creative breakthroughs, correcting disciplinary oversights and blind spots, addressing topics that fall through the cracks because they do not fit into any established discipline, and addressing complex real-world problems that require us to move beyond the tunnel vision of experts (1997, 204–8). Klein concludes her book by saying that "cutting across all these theories is one recurring idea. Interdisciplinarity is a means of solving problems and answering questions that cannot be satisfactorily addressed using single methods or approaches" (1990, 196).

Such general praise for interdisciplinary approaches is all well and good, but I would like to focus on the need for interdisciplinarity in investigating scientific knowledge, that is, within the field of science studies where I locate this project. And here we find a welter of recent declarations that in order to understand scientific knowledge we must use more than one disciplinary approach. The argument is often based on the claim that science is an activity of great complexity; as Ronald Giere puts it, science is "at least as complex as the reality it investigates. This great complexity implies, I think, that it is impossible to obtain an adequate picture of science from any one disciplinary perspective. Different perspectives highlight different aspects while ignoring others" (1999, 28). Two of these different aspects are the sociological feature of competing interests and the epistemological feature of experimental data. Giere conceives of these as complementary factors in accounts of scientific theory choice: "In some cases experimental data may strongly influence theory choice; in other cases political commitments or professional interests might be dominant. Most cases are mixed" (1999, 61). Giere advocates a multidisciplinary approach, saying that those with different perspectives in science studies need to collaborate and to integrate their approaches (63). But there are different ways of integrating perspectives, and we need to be clear that what is called for is not always the creation of new, hybrid disciplines or grand unified theories.

Alison Wylie has spoken of a growing consensus within science studies that "each of the existing science studies disciplines is inherently limited, taken on its own. Indeed, given the complex and multi-dimensional nature of scientific enterprises—a feature of science that is inescapable when you attend to its details—it is simply implausible that the sciences could be effectively understood in strictly philosophical, or sociological, or historical terms" (1995, 394). And this point finds further support from the chemist Henry Bauer (1990, 113), the philosopher David Stump (1992, 458–59), and the sociologist Andrew Pickering (as cited in Wylie 1995). Despite this impressive consensus, one may still ask: Why does complexity require interdisciplinarity? After all, chaos theory itself has shown us that complicated behavior may be the result of a simple underlying dynamic. But the complexity of science lies not only in its behavior—its "output" of successive theories, for instance—but also in the multiplicity of factors at work in its dynamic processes and in the multiplicity of questions that can be asked about them. The sheer complexity of an object of inquiry, which may be a system with a multiplicity of relevant factors, does not in and of itself require a plurality of interacting disciplinary approaches. If each discipline simply focused on one aspect of the system being studied, their separate investigations could be straightforwardly added together for a comprehensive account. But when different disciplines ask different kinds of questions, animated by different concerns and interests, their approaches will not simply attend each to its own relevant features. I return to this question later in the course of articulating an image of multidisciplinary science studies that goes beyond the conception of multiple perspectives.

If, as I have argued here, we should use more than one disciplinary approach in investigating scientific knowledge, how much more important might it be when we broaden our scope to include a consideration of how scientific knowledge is used in other fields? For example, when legal scholar Glenn Harlan Reynolds (1991) claims that the pattern of decisions by the United States Supreme Court should be thought of as a strange attractor, we may ask a number of different questions. We may want to know why he has chosen to borrow concepts from chaos theory, or what effect his argument has on his audience, or what implications this argument has for the practice of law. We may even want to know whether and in what respect the pattern of Court decisions is in fact like a strange attractor. To answer these questions, we need the resources of such disciplines or subdisciplines as the rhetoric of inquiry, the sociology of knowledge, the philosophy of law, and the linguistics of metaphor. These different approaches interact and inform one another but cannot be integrated into one general scheme for the examination of borrowing, in part because they examine different objects of study: legal scholarship is a practice that generates knowledge, but it is also a discourse that generates money, power, prestige, and policy recommendations. I return to the issue of the incompatibility of disciplinary approaches in the course of developing a new image for interdisciplinarity.

Not everyone shares an enthusiasm about interdisciplinarity and multiple approaches. For one thing, interdisciplinarity can become indiscriminate or exclusive when it becomes an end in itself.[4] It is important to recognize that fruitful connections between disciplines depend on the existence of methods of inquiry developed within particular disciplinary contexts. So interdisciplinarity can be problematic if it is all that we do and is seen as more important than the pursuit of any particular discipline. But that is not a position I advocate. Instead, disciplinary pluralism affirms the use of multiple approaches when appropriate, while recognizing that sometimes a single approach is called for. More is not necessarily better. David Stump counsels that, as with science, we should test our methods and see which ones work: some will probably be unproductive for some purposes. Different disciplines have different perspectives to offer and may all be useful at different times for different purposes (1992, 459).

A second concern about interdisciplinarity, the danger of dilettantism, is raised by Nissani (1997, 212) and Bauer (1990, 113), among others. After all, there is a reason why serious fields of inquiry are called "disciplines"—their practice requires time, dedication, and indeed discipline. The demands of rigorous and specialized scholarship make it exceedingly difficult to engage responsibly with more than one discipline. Dabblers may easily be misled by superficial resemblances when they are not acquainted with the technical details wherein so much of the real effort lies. Joe Moran admits, "it could be argued that, because they are relatively new and exploratory, interdisciplinary ways of thinking have a tendency to be more disorganized, error-prone and incomplete than established forms of knowledge" (2002, 184). But he suggests that such messiness may be a necessary price to pay. In response to these concerns, I suggest that there is nothing about interdisciplinarity that necessitates a lack of seriousness. Instead of criticizing or defending interdisciplinary work at the general level, we should examine particular instances of interdisciplinary work and ask whether they are superficial dabbling. Perhaps interdisciplinary efforts have a greater percentage of superficiality, but I have found no systematic evidence of this. Surely narrow monodisciplinary efforts are not free from the risk of superficiality themselves. In the end, the difficulty of becoming conversant with more than one field can be outweighed by the new insights that can be gained.

This point parallels a more general theme of this project. Interdisciplinarity is neither always good nor always bad. It is neither automatically an end in itself, worth pursuing in all cases to the exclusion of disciplinary effort, nor is it always mere dilettantism. It can be useful or useless, as the case may be, and this is true at the level of borrowing as well as at the meta-level of how to go about examining borrowing. "Dilettante" is simply a term of abuse for bad scholarship that deviates from accepted turf boundaries. Bad scholarship is the problem; interdisciplinarity itself carries no special risk. Or rather, if it does carry a special risk, this is because the existing structures for rigorous criticism of interdisciplinary work are presently weaker in many areas. And it is precisely this weakness that my project seeks to remedy, by creating some tools for the rigorous critique of cross-disciplinary work.

Disciplinary Pluralism as Cross-Training

Debates about interdisciplinarity often use metaphorical imagery, so it serves us well to examine some of these metaphors and seek one that helps to highlight a pluralistic approach. As I argue in chapter 5, metaphors matter for more than just explicating or illustrating a point, because they highlight or conceal particular features that are important. To clarify how disciplinary pluralism contrasts with other accounts of interdisciplinarity, I propose the metaphor of cross-training in sports. The idea behind cross-training is that one can improve one's performance in a chosen sport by training in another. In the 1980s it was not unheard of for wide receivers to take ballet lessons to learn grace and agility. I have heard that Kareem Abdul-Jabbar took up karate to improve his performance at basketball, and that figure skater Elvis Stojko studied kung fu to improve his speed and balance. Hockey players may practice gymnastics to improve agility, and swimmers may study yoga to improve concentration. Different sports help athletes develop different skills, and an exclusive focus on one sport may even become counterproductive. Any number of different sports may contribute to one's goals, although one does not have to practice every sport, and indeed, cross-training does not necessarily work for everyone.[5]

The contrast between the image of cross-training and other metaphors for interdisciplinarity helps articulate the nature of disciplinary pluralism. The other metaphorical images to be considered are nations, tiles, and languages. Julie Thompson Klein offers a comprehensive survey of metaphors for disciplinary relationships and points out that the dominant image is drawn from geopolitics, with disciplines conceived of as nations. Hence, disciplines have boundaries, borders, and frontiers; turf, territory, and no-man's-land; expeditions and migration; balkanization, protectionism, and autonomy; and nationalism, tribalism, ethnocentrism, and imperialism (Klein 1990, 77; see also Gieryn 1999). She also mentions the images of fish scales, honeycombs, and perspective slices of a solid, and these are what I refer to as "tiles"—disciplines conceived of as separate pieces of knowledge that map onto contiguous parts of the world (Klein 1990, 81–82). The third main cluster of images I discuss involves the metaphor of disciplines as languages, each describing the world differently.[6]

Looking first at disciplines as nations, we find that Donald T. Campbell provides an early example of this image when he identifies the "ethnocentrism of disciplines" as "the symptoms of tribalism or nationalism or ingroup partisanship in the internal and external relations of university departments, national scientific organizations, and academic disciplines" (1969, 328; see also Kant [1787] 1929, 18 Bviii). A pluralist can make sense of the nation metaphor by saying that certain questions, like adjustments in income tax rates, are matters of internal politics within a sovereign nation. This example corresponds to the idea that monodisciplinary approaches are sometimes appropriate. Other matters require international cooperation, as when a lake straddles a border. Still other matters may require the forging of alliances or the creation of transnational bodies such as the United Nations, and there are even some people who call for dissolving all nations into bodies such as the European Union or a world government. These situations correspond to varying forms of cross-, multi-, and transdisciplinary endeavors, each appropriate for different types of inquiry.[7] However, the nation metaphor paints a static picture, where disciplines are seen as organized in space rather than as ongoing human activities. Because of our contemporary geopolitical conception of the nation-state, the image of disciplines as nations highlights competing social structures ("turf") and objects of study ("domains"). It also calls attention to the role of disciplines in regulating discourse ("policing"). But this metaphor diverts attention away from disciplines as collections of cognitive tools for active investigation. The image of cross-training avoids this limitation.

We can find the image of disciplines as "tiles" in Ronald Giere's argument for multiple disciplinary perspectives in science studies, where he says that "the only adequate overall picture will be collages of pictures from various perspectives" (1999, 28). Speaking of disciplinary approaches as "perspectives" or "views" conjures the image of a preexisting object with several aspects. Different disciplines may examine different aspects, or they may examine one and the same aspect from different directions. Sociology, for example, will look at funding patterns and networks of training, but rhetoric will not. Yet this characterization of multidisciplinarity works best for situations in which a number of independent and clearly demarcated causal factors are at work, with each perspective identifying one factor. Certainly the tiles may overlap in some areas, or have porous boundaries, but the conceptions of partial perspectives or contributing factors promise an easy compatibility. In more tangled situations, multiple perspectives may be necessary, but the views provided may not all fit together or lie flat. It is not clear that we can parcel things out tidily, for example, attributing 60 percent of an episode of theory change to the evidence and 40 percent to social interests.[8] Again, the tiling image of multiple visual perspectives limits us to a static picture, whereas the metaphor of cross-training encourages us to think of active endeavors that may not be able to be practiced simultaneously. You cannot do yoga while swimming (at least, not for long), and it is similarly difficult to maintain a sociologist's methodological relativism while engaging in epistemological evaluation. And, as I discuss further in the next chapter, rhetoric and epistemology both look at arguments but engage with them in very different ways.

Campbell's image of comprehensive knowledge involves a fish-scale pattern: "a continuous texture of narrow specialties which overlap with other narrow specialties. Due to the ethnocentrism of disciplines, what we get instead is a redundant piling up of highly similar specialties, leaving interdisciplinary gaps" (1969, 328). This description evokes an ideal image of slightly overlapping shapes, tiling the terrain much more effectively than our current situation, where clusters of highly overlapping fields leave broad spaces between clusters. But such an image suggests that each piece of the world has one discipline (or at most a small number of cooperating disciplines) appropriate for studying it. A thoroughgoing disciplinary pluralism goes deeper and suggests that sometimes the perspectives do not fit nicely together on the same plane: they overlap or conflict or cannot both be held at the same time, and yet both are needed to understand the phenomenon. This disciplinary pluralism shares much in common with the philosophical pluralism advocated by Robert Nozick, which allows for a multiplicity of admissible views. The multiplicity does not entail relativism, however, because not all views are admissible, and the admissible views can be ranked on their merits. As Nozick says, "the first ranked view is not completely adequate all by itself; what it omits or distorts or puts out of focus cannot be added compatibly, but must be brought out and highlighted by another incompatible view, itself (even more) inadequate alone" (1981, 22).

The third image to consider is that of disciplines as languages. Chemist Henry Bauer uses this metaphor to point out that just as languages have different grammars and not merely different vocabularies, disciplines not only deal with different facts but have different ways of doing things (1990, 112). So this image, unlike that of disciplines as nations or tiles, draws attention to disciplines as modes of human activity, especially if we view a language as a tool and not merely a collection of vocabulary and grammar. Bauer suggests that some languages are good for some purposes and not others and uses this point to argue for the need for multidisciplinary approaches in science studies (113). Perhaps some languages are especially suited to certain tasks—for instance, some claim that the ability to easily coin new terms by concatenation facilitates philosophizing in German. But a number of mismatches detract from the usefulness of the language metaphor for conceiving of interdisciplinarity. For instance, it is not clear that there are many situations outside of linguistics in which one needs to know a number of different languages in order to solve a problem or express an idea.

Different metaphorical images highlight different features of interdisciplinarity, and I advocate metaphorical pluralism as well. So, despite its limitations, we should not dismiss the metaphor of disciplines as languages too quickly. Indeed, some interesting research on cross-disciplinary work has made use of this metaphor, especially in terms of pidgin languages. Peter Galison describes the way physicists and mathematicians with different specialties created a pidgin language to enable them to communicate about the Monte Carlo procedure.[9] Eventually, the pidgin became a full-fledged Creole, a language of its own (Galison 1996, 152–53; see also Fuller 1996, 171). The image of languages helps us avoid static spatial pictures of interdisciplinary relationships and provides a valuable metaphor for the creation of intermediate disciplines. But the cross-training metaphor not only highlights activity but helps illustrate the fruitful interactions between different activities. Although we sometimes borrow words, we rarely find it helpful to use several languages to try to say the same thing.

Challenges to Disciplinary Pluralism: Resisting Collapse and Insulation

A pluralistic approach offers many benefits. Not the least of these is that pluralism conceived as cross-training opposes both the notion that there is one true, best sport for everyone and also the notion that each sport is intrinsically separate and has nothing to do with any other. These two points of view find expression in the metaphor of languages as the ideal of a single universal, Adamic, perfect language and the view of absolute untranslatability. In the idiom of nations, these positions go by the familiar names of imperialism and isolationism, and in the metaphor of disciplines as tiles, they correspond to the views that one tile is enough to comprehend the whole universe and that each tile is strictly separate from its neighbors, with no overlap or vagueness at the boundary. Although the metaphor of cross-training provides a salutary image for displaying the virtues of disciplinary pluralism, unfortunately the language of sport lacks convenient terms for the vices corresponding to imperialism and isolationism. Yet we can imagine a fanatical gymnast insisting that his or her sport is the foundation of all others and subsumes them, or a separatist golfer insisting on a pure and single-minded devotion to the game.

Dealing with these views in turn, let us first confront the notion that one discipline is ultimately sufficient for all forms of inquiry. This disciplinary imperialism may take the form of an extreme reductionism that Nancy Cartwright calls "fundamentalism," a claim that all truths are ultimately truths of physics (1999, 25). Or it may appear in the guise of grandiose claims that all the social sciences and the humanities can ultimately be reconceived as branches of biology (Wilson 1980, 271). Pluralism does not rule out the possibility of sometimes reducing one field of study to another, but despite the occasional real successes of reductionist approaches, interactions between disciplines encompass a much greater variety than simply absorption into one ultimate unifying field of inquiry. Darden and Maull point this out for scientific disciplines (1977, 60), and Giere insists that in science studies no single theoretical account can be adequate (1999, 63). But the threat of disciplinary imperialism comes not only from the sciences. Kenneth Knoespel identifies a danger from theorizing humanists who "make claims to account for the sciences" and evince "the arrogance present in certain literary theory that would absorb virtually everything in a handy epistemological schema" (1991, 119). Celebrated literary theorist Stanley Fish correctly diagnoses an imperialist tendency in every discipline, and he derides the notion that literary or historical (or, indeed, philosophical) expertise can serve as a generalized form of wisdom. He connects the imperialist temptation to "the quite ordinary tendency to believe that the skills attending one's own practice are indispensable for any and all practices" (1995, 89; see also Dupré 2001, 16). Fish has also identified a strand of cultural studies that envisions itself as an antidisciplinary synthesis that escapes all strictures of professional authority (1994, 78). And the sociologist Andrew Pickering similarly speaks of the "collapse" of all disciplines into an antidisciplinary synthesis for science studies (1995, 416).

So imperialist transdisciplinary talk can come both from the direction of universalizing reductionism and from constructivist critics of disciplinary boundaries. But just because the disciplines are historically contingent, multiple, fractured, and blurry does not make them unreal or mean that they all ought to collapse their domains and methods into one (see Fish 1994, 240). Although I noted earlier that disciplinary research contains elements of openness, plurality, and even opportunism, nonetheless rough distinctions between objects of study and techniques of inquiry apply. Even if the socially constructed boundaries between academic departments were to wither away, why should we think that one kind of training, method, or approach will always suffice? In philosophy, for example, translating arguments into first-order logic sometimes works, but surely it will not solve every problem we face. Pluralism rejects the assumption of monodisciplinarity, whether it wears the guise of universalizing reductionism or of a totalizing collapse into one undifferentiated endeavor.

Of course, it is possible that I am committing some disciplinary imperialism of my own, swaggering in across genres and blithely assuming that philosophical standards of argumentative analysis are appropriate to apply to literary interpretations or law reviews. And I admit that I claim no special expertise in adjudicating whether a reading of a particular novel is particularly illuminating. It may be that essays on literature are meant to be seen as displays of interpretive virtuosity, or as tentative speculations, or as personal reflections. But when authors say that one thing is like another, I feel invited to step in and say that the similarity is not so strong. When authors present theses about the nature of literature, or state that one theoretical view is incompatible with another, it is perfectly legitimate to read these as claims that ought to be supported by reasons. Similarly, the analysis of economic data and the reading of specific legal cases often turn on precise technical details inaccessible to those (including me) without the relevant professional expertise. But general theoretical arguments in economic policy or jurisprudence cannot be so easily insulated from criticism.

Opponents of disciplinary imperialism who wish to defend the contingent integrity of disciplines face the opposite temptation of falling into isolationism—the view that each discipline is utterly separate. Ronald Shusterman suggests such a strategy as a modus vivendi for resolving the "two cultures" controversy, claiming that art and science are "incommensurable" and so any conflict between scientists and humanists has no point: "A better form of pragmatism would tell us that the different discourses have different jobs to do, and leave it at that" (1998, 134). This solution holds that science deals with the external world, whereas literature and philosophy deal with the inner world, and each side should leave the other alone. But some have called for isolation even within the humanities. Moran cites F. R. Leavis as calling for the separation of literature from philosophy to prevent "blunting of edge, blurring of focus and muddled misdirection of attention: consequences of queering one discipline with the habits of another" (2002, 31). Those who want to be good at football, this view seems to claim, have nothing to gain from pursuing boxing or ballet. But in place of such purity and containment, pluralism holds out the possibility of mutually helpful cooperation.

Stanley Fish's emphasis on the unique distinctiveness of disciplines leads him to deny the possibility of such interaction, however. He says that "whenever there is an apparent *rapprochement* or relationship of co-operation between projects, it will be the case either that one is anxiously trading on the prestige and vocabulary of the other or that one has swallowed the other." According to Fish, "each discipline has its own job" and its own vocabulary, "and when one discipline borrows from another—as the critic may borrow from the historian—the borrowed material is instantaneously transformed into grist for the appropriator's mill" (1995, 83). We may grant that material borrowed will be thus transformed, but it does not follow that disciplinary practices are insulated from the possibility of genuine communication and transformative contact. The borrowing discipline may be transformed as well. Although knowledge can only be made within the context of a community of inquiry, Fish misconceives knowledge as being so utterly fixed by that disciplinary context that communities can absorb the products of other disciplines only by reinscribing them according to their own rules.

The image of cross-training can go some way toward illustrating the benefits of interaction between practices, but no one metaphor can do all the work required. Moti Nissani recognizes that the blending of disciplines is not always called for when he proposes a measure of interdisciplinary integration using the metaphor of disciplines as fruits: "The various fruits can be served side by side, they can be chopped up and served as a fruit salad, or they can be finely blended so that the distinctive flavor of each is no longer recognizable, yielding instead the delectable experience of the smoothie. Note that the amalgamation quotient says nothing about quality: in some circumstances, a plain mango will surpass all the smoothies in the world; in others, only a fruit salad will do" (1995, 125). An additional possibility would be the macédoine, in which the pieces are combined with a syrup and their flavors encouraged to blend while remaining distinct. Such a dish serves as a metaphor for disciplines that retain their integrity while entering into a fruitful interaction.

Pluralism asserts that there may not be just one proper relationship between disciplines: not imperialistic fundamentalism, nor collapse into transdisciplinary synthesis, nor isolation into incommensurable magisteria. Although each of these may be called for in some circumstances, in some circumstances cross-disciplinary transfer and interaction will prove fruitful and appropriate. Giere makes a plea for such interaction when he says that science studies "must draw on knowledge from many disciplines, including some of the sciences it studies" (1999, 29). In philosophical endeavors this view is called "naturalism," and challenges to naturalism are among the main barriers to a pluralistic approach in science studies. Another barrier, related to objections to naturalism, is the traditional philosophical distinction between the context of discovery and the context of justification. This distinction seeks to isolate epistemology from empirical disciplines such as history and sociology, so a defense of disciplinary pluralism must address this distinction head on. In the remainder of this chapter I take up these isolationist challenges by undermining the "two-contexts" distinction and by defending a naturalistic pluralism.

Overcoming the Two-Contexts Distinction

The distinction between the context of discovery and the context of justification is meant to capture two different aspects of knowledge-making processes. The context of discovery is supposed to concern the way a person generates an idea, and it is suggested as the legitimate area of investigation for psychologists and historians. The context of justification is meant to capture an entirely different aspect, namely, the way an idea (wherever it came from) becomes rightfully certified as knowledge. The right to be so certified would not depend on the actual justificatory practices of the scientific community but would instead depend on a particular logical relationship between the idea and the evidence that supports it. As a result, the context of justification was held to be the exclusive province of epistemologists and philosophers of science. After all, Karl Popper and the logical empiricists generally thought that philosophy should be logical analysis of the distinctive way that scientific theories were supported (or refuted) by evidence. Discovery was viewed as irrelevant, unimportant, and not amenable to logical reconstruction.[10]

For the logical empiricists and Popperians, the distinction gave philosophy of science a foundational difference that served to demarcate it from psychology, history, and sociology. This distinction played a crucial role in isolating philosophy of science from other disciplines that also investigated scientific knowledge-production. Philosophers in favor of collaboration with empirical disciplines, such as those whose historicist approach implied that other disciplines had something important to say to epistemological concerns, found their opponents wielding the distinction in order to claim that they had "simply missed the point of what philosophy of science is all about" (Hoyningen-Huene 1987, 501). Thus a defense of disciplinary pluralism in the study of scientific knowledge must confront this distinction.

In fact, the distinction poses a sharp threat to the entire project of an epistemological examination of borrowed knowledge. The practice of borrowing, after all, seems to belong solely to the context of discovery. If the generation of ideas at the first stage of knowledge production utterly lacks epistemological interest, then the only thing left for philosophy to do when faced with borrowed knowledge is to ask of any particular thesis whether it relates in the proper way to the evidence. The fact that a thesis was formulated in emulation of some result from another discipline would be strictly irrelevant to its status as knowledge, so borrowing could serve merely as a source of inspiration or glamour. Because inspiration belongs solely to the context of discovery and glamour plays a fairly minor role in logical relationships of justification, nothing epistemologically interesting can be said about borrowing. Or, as one philosopher has put it, this project has merely sociological interest.

To meet this challenge, I identify and address three dimensions of the distinction between the context of discovery and the context of justification: we can see a distinction being made between generation and testing, between the empirical and the logical, and between the descriptive and the normative. Arguments against each of these dimensions of the two-contexts distinction contribute to the case for disciplinary pluralism. First, we see that the two phases of discovery and justification are blurry and interpenetrating, with an intermediate stage of pursuit. Next, we begin to see the generation of ideas as a reasoned process, with normative and justificatory aspects. The bright line between the empirical and the logical begins to blur. Finally, we see that justification itself involves practical and social aspects in the testing and acceptance of theories, and this understanding leads us to a further discussion of the distinction between the descriptive and the normative.

Clarifying the Distinction
What exactly is the distinction in question? Hans Reichenbach originated the current terminology when he sought to separate two tasks in the examination of inductive reasoning. On one side of the line would be creation of an idealized reconstruction of the process of evaluating such reasoning—the context of justification. On the other side would be investigation of the actual thinking processes involved in the origin of new scientific hypotheses—the context of discovery (Reichenbach 1938). But the two-contexts distinction in a broader sense is actually quite

widespread and long standing, appearing in the work of Popper, Carnap, Husserl, Frege, Whewell, Herschel, and even Kant (Hoyningen-Huene 1987, 503). The two-contexts distinction encompasses a number of distinctions that align along three broadly conceived dimensions. First, we find a temporal distinction between two different stages in knowledge production: the process of theory creation and the process of theory testing. Second, there is a distinction between the logical and the empirical, where empirical disciplines such as history, psychology, and sociology are separated from philosophy conceived as the logical explication of the basis of scientific theories. Along this dimension, epistemology would be solely concerned with the context of justification, using modern formal logic to examine the timeless and context-free relationships between statements. Thus, it would have nothing to learn from empirical inquiry. Conversely, the context of discovery contains nondiscursive moments of inspiration—mental episodes of the "aha" experience, which are nonlogical and hence utterly separate from issues of justification. Third, the distinction is used to contrast descriptive questions (about how ideas *are* generated) from normative questions (about the methods that *ought* to be used to justify said ideas). With this sharp line drawn between the descriptive and the normative, the disciplines of psychology, history, and sociology find themselves confined to purely descriptive inquiry, and the process of discovery is held to have no normative dimension.[11]

The two-context distinction as it has usually appeared applies to scientific knowledge, but for the purposes of investigating the phenomenon of borrowing we can consider that the distinction might apply outside of the natural sciences. After all, researchers in the social sciences and the humanities generate ideas and attempt to justify them as well. The reader will also note that I have been speaking of the distinction in the past tense. Indeed, in 1987 Hoyningen-Huene pointed out that interest in the distinction had faded away and that empirical inquiries were now taken more seriously in philosophy of science (1987, 502). Nonetheless, it is worthwhile to sketch briefly some of the reasons for its passing, in order to see how they support a pluralistic approach to science studies and undercut the assertion that borrowed knowledge is of no epistemological interest.

The Passing of the Temporal Distinction

Looking first at the failure of the distinction along temporal lines, we find that a consensus has emerged that different phases of inquiry overlap and blend, with no clear delineation in time. Susan Haack suggests that the time has come for us to recognize that "the exploration and articulation, testing, modification, presentation of an initial vague idea can take place together, or in an up-and-back order" (1998, 80). So the older picture of two isolated phases of inquiry has been largely supplanted by a scheme with three overlapping stages: a *generation* stage, in which researchers produce an idea worth a second look; the *pursuit* stage, in which researchers engage in preliminary evaluation of the idea, assessing its plausibility, comparing it with alternatives, and elaborating it, which leads to active consideration of the idea; and the *acceptance* stage, in which the community certifies the idea as knowledge (Nickles 2000, 90; 1980, 10; Hoyningen-Huene 1987, 507–8). The recently introduced context of pursuit includes articulation, modification, and preliminary evaluation, and thus holds significant epistemological interest.

The in-between character of the context of pursuit calls attention to the shifting boundary between generation and acceptance, for the same scientific work can be viewed as part of the context of discovery or the context of justification, depending on the historical framework from which one approaches it (see, e.g., Franklin 1993, 107). But does the failure of a strict temporal segregation between discovery and justification have anything to do with the deeper distinction between empirical and logical modes of inquiry? Perhaps the admission of a hybrid, intermediate context of pursuit merely shows that there is a third stage between a fully nonrational generation and a solely justificationary phase of testing and acceptance. So we must examine the stage of idea generation itself to see whether it is genuinely devoid of epistemological interest.

A tradition going back to Norwood Russel Hanson's 1958 book, *Patterns of Discovery*, claims that discovery itself has a logic. But the search for a general, content-neutral logic of discovery, as opposed to various heuristics for specific problems, sets the sights too high. A focus on reasoning (as opposed to logic, narrowly construed) makes it easier to see discovery as philosophically important (Nickles 1980, 25–26). The crucial question is whether the initial generation of ideas as possibilities for consideration has something to do with justification and epistemology. For, as Stephen Toulmin has pointed out, "to regard something as being a possibility at all is, among other things, to be prepared to spend *some* time on the evidence or backing bearing for it or against it" (1958, 18). It is not a matter of saying that the circumstances of the generation or the generator (their origin in a dream, or the gender or ethnicity of the scientist) must themselves count in favor of a theory. But not every fleeting fancy is even considered an idea worth remembering at all; the generation of new ideas always includes some evaluation because the very fact that an idea appears as a possible solution to a problem means that it satisfies some of the constraints on a solution, which itself has something to do with justification (Nickles 1980, 13, 22).

So we find that the generation stage is not as simple as we might expect, although some science textbooks continue to describe hypothesis generation as "guessing." As Nickles argues, contemporary philosophers sought to counteract a naïve account that scientific hypotheses arise through simplistic generalizations about tediously compiled facts. To combat this Baconian inductivist dogma, they repeated anecdotes about Kekulé envisioning the structure of the benzene ring in a dream. But the difficulty of giving a rich account of the origin of scientific discoveries does not mean we have to lapse into a view that they all come from instantaneous flashes of intuition. If the generation stage is utterly devoid of methodological and epistemological interest, then all methods for coming up with new hypotheses are equally good. And if the generation process is completely irrational or unanalyzable, then how can historians make it intelligible? Instead, we should see generation itself as a reasoned process (Nickles 1980, 17, 29–31, 40). Indeed, the process of generating new ideas receives some systematic attention from rhetorical theorists, whose analyses of the methods of "invention" play a role in chapters to come.

The Empirical and the Logical
So we have seen that the pursuit of ideas involves some features connected with justification. We have seen that generation itself is not an irrational, unanalyzable realm with no elements of epistemological interest. Looking now at the third phase, of testing and acceptance, we can see that it is neither narrowly logical nor separate from social concerns. The idea that philosophy of science should elucidate a logic of science with certain and context-independent rules has fallen by the wayside in light of the recognition that most interesting scientific reasoning is content specific. One can find instances of patterns of inference such as *modus ponens*, if that is what is desired, but they will not tell us much about science (Nickles 1980, 16).

Indeed, in considering the processes of justification and acceptance, we also confront the distinction between the empirical and the logical. For philosophy of science has come to recognize that the actual processes of scientific justification need to be investigated in all their empirical actuality, not merely as idealized patterns of logical inference. This recognition has finally made good on Stephen Toulmin's exhortation that reasoned arguments that justify ideas extend wider than strictly deductive arguments in the canons of modern formal logic, so "rational demonstration is not a suitable subject for a timeless, axiomatic science" (1958, 147). If we are interested in the justification of knowledge, we cannot look only for universal criteria that decide the strength of any argument, but we must expand the analysis of argumentation to include the empirical details of particular situations.

Furthermore, Kuhn (1977) and others have argued that theory choice involves sociological and psychological factors. If what counts as a good scientific reason, one that actually justifies a choice, depends in part on features of the relevant knowledge community, then the epistemological examination of justification can never be separated from empirical investigation into the actual practices of justification (Hoyningen-Huene 1987, 509). Thus, attention to the final stage of knowledge production, acceptance, undercuts the supposed distinction between purely logical issues of justification and purely empirical questions of actual scientific practice.

The Descriptive and the Normative

The final blow to the distinction between context of discovery and context of justification is a challenge to the purported split between descriptive and normative enterprises. This challenge is posed by naturalized epistemology and naturalistic philosophy of science, which insist that empirical disciplines are relevant to epistemology. As Nickles points out, a methodology that treats the processes of generation and pursuit would have a normative and critical component, as all methodologies must. So the distinction between these earlier phases and the later phase of acceptance will not map neatly onto a descriptive/normative distinction. Neither will the distinction provide a sharp separation between the genuinely philosophical and the nonphilosophical (Nickles 1980, 20). We need to have a methodology that includes all phases of the knowledge-making process, one that would be both descriptive and normative.

Hoyningen-Huene has taken up this issue in some detail. He states that the distinction between the descriptive (or, as he calls it, the "factual") and the normative is at the core of the two-contexts distinction, and it is a difference in perspective, not in subject matter. On the one hand we have description of processes and facts, and on the other we find a concern for the appraisal and evaluation of whether and how cognitive claims are justified. But a distinction between factual and normative does not imply that they are isolated and have nothing to say to each other; after all, we can describe norms, and we must use norms to make descriptions. In the end, the descriptive and the normative are two kinds of questions we can ask about a claim, from two different perspectives

(Hoyningen-Huene 1987, 511). Toulmin makes a similar point that the question "How do you know?" sometimes calls for a logical answer in terms of evidence and justification, and sometimes for a biographical answer in terms of personal experiences and one's train of thinking. Sometimes it is not a question of whether or why people think as they do, but whether their argument is up to standard and deserves to be accepted (Toulmin 1958, 215–17).

Hoyningen-Huene claims that attacks on the two-contexts distinction do not erase the difference between the factual and the normative, which is a valid and important distinction. Indeed, to simply collapse the difference between discovery and justification could lead to disastrous situations in which ideas are rejected simply because of who came up with them. But the challenges discussed earlier call into question the conflation of the distinction between the factual and the normative with the distinction between types of processes, or between the logical and the empirical, or between philosophy and empirical disciplines. If, as Nickles suggests, we give up a sharp dichotomy between the genuinely philosophical and the nonphilosophical, it should not be because there is no difference between the questions asked by different disciplines or between descriptive and normative questions. Rather, it should be because methodology and epistemology must look at both kinds of questions and should consider the descriptive as informing the normative and vice versa. As Helen Longino puts it, "a philosophical theory of knowledge is neither purely descriptive nor purely prescriptive" (2002, 10). It involves the descriptive task of characterizing knowledge producers and the processes they engage in, but it also requires an account of the normative conditions that must be met for certain beliefs, theories, and the like to correctly count as knowledge. Therefore, nonphilosophical inquiry can be crucially relevant to philosophical questions, and this useful interaction is precisely what disciplinary pluralism calls for. Yet, because this pluralistic approach can to some extent be identified with naturalistic approaches in the philosophy of science, we must confront head-on one of the challenges faced by naturalism: the charge that it cannot in fact reach normative conclusions and must therefore fail as an epistemological enterprise.

Normative Naturalism

A disciplinary pluralist will be perfectly happy to accept that the descriptive and the normative are indeed different approaches with different questions to ask. Neither approach should be dismissed, disrespected, absorbed, or eliminated by the other. Yet although these approaches are *different*, they need not be kept strictly *separate*. Descriptive results can contribute to evaluation, and normative inquiry has something to learn from empirical disciplines. These different projects can sometimes influence and inform each other in helpful ways (including, it is hoped, the analysis of borrowed knowledge that is to come). One name for this pluralistic coexistence is normative naturalism.

Naturalism

Naturalism in epistemology (as well as in philosophy of science and science studies) means something quite different from naturalism in literature or even in ethics. Epistemological naturalists usually trace their lineage back to Quine's "Epistemology Naturalized," in which he states that "we are after an understanding of science as an institution or process in the world, and we do not intend that understanding to be any better than the science which is its object" (1994, 26). Many proponents of naturalism, including myself, disagree with many of Quine's views in this founding document. But we can see here two of the crucial features shared by most forms of naturalism: an opposition to supernaturalism and a belief that epistemological inquiry is continuous with other kinds of inquiry (and specifically, scientific inquiry). Clarifying these two features will enable us to see how naturalistic approaches connect the descriptive and the normative in the spirit of disciplinary pluralism.

Let us start by noting how naturalism stands in contrast to supernaturalism. At its root, naturalism says that our knowledge-making efforts are part of this world, the world we experience. As such, our knowledge is always subject to review when we learn more about this world, and we need not go looking for knowledge that is ultimately independent of our interactions with this world (Brown 1988, 64). It is important to notice that this version of naturalism has an epistemological, as opposed to a metaphysical, flavor. It does not pretend to rule out causes that outstrip our current physical sciences but simply encourages us to retain contact with the actual ways we go about seeking knowledge. Giere calls this a "methodological naturalism," which offers suggestions about how to go about gaining knowledge, as opposed to a "theoretical naturalism," which advances metaphysical theses about nature itself (1999, 77).

The second important feature of naturalism, continuity with the sciences, receives an especially contentious expression from Larry Laudan. For Laudan, naturalism holds that "the claims of philosophy are to be adjudicated in the same ways that we adjudicate claims in other walks of life, such as science, common sense and the law. More specifically, epistemic naturalism is a meta-epistemological thesis: it holds that the theory of knowledge is continuous with other sorts of theories about how the natural world is constituted," and that the theories of naturalistic epistemology are subject to "precisely the same strategies of adjudication that we bring to bear on the assessment of theories within science or common sense" (Laudan 1990, 44–45). Such bold language unfortunately downplays the fact that different fields adjudicate claims differently. Better to think of naturalistic enterprises not as precisely the *same* as other disciplines but *continuous* with them in the sense of making contact across disciplinary boundaries.

The Descriptive and the Normative

The question of how the descriptive can be relevant to the normative is usually framed as a challenge to naturalized epistemology, to which its defenders reply in a spirit of either aggression or reconciliation. The most aggressive version of naturalized epistemology holds that normative epistemological questions must be replaced by empirical questions because there is nothing left for epistemology to do. The milder response, more aligned with pluralism and the primary focus of this section, holds that epistemology asks distinctive questions, different from those pursued by empirical inquiry, but that these fields can interact profitably.

Unlike the milder response, the aggressive version views naturalism as implying the replacement of normative epistemology by scientific inquiry. This conception goes back to the birth of naturalized epistemology in Quine's original article from 1969, in which he advocated jettisoning philosophical attempts to justify our knowledge and instead "settling" for psychology, seeking only to describe the way observations lead to scientific knowledge (1994, 20). Jaegwon Kim points out that when Quine urges us to eliminate the normative project, we face the question of whether we are even doing epistemology anymore. Kim acknowledges that most naturalists are not Quinean eliminativists and still have a normative project (1994, 43–48), but the problem of normativity is often used as a criticism against naturalistic projects. Hoyningen-Huene has defended the more moderate naturalistic approaches by pointing out that claiming the relevance of empirical disciplines is far different than claiming that epistemology is replaceable by them (1987, 510).

Another kind of elimination of the normative comes not from bold claims of replacement by the psychological sciences, but from a renunciation of the normative project in favor of empirical description of the social practices of belief acceptance. Barry Barnes and David Bloor explicitly embrace this relativist approach when they focus solely on investigating the causes of belief becoming accepted, while holding at arm's length the evaluation of its correctness (1982, 23). Yet bold disavowals of normative appraisal are seldom maintained in practice, and this is all for the better. As we see in the next chapter, some of those who loudly proclaim that they have jettisoned all ambition to pass judgment nonetheless make judgments and make them with good reasons. Social scientists who forswear all normative claims cannot help us with our task of evaluating different instances of borrowed knowledge and may in fact be aspiring to a distorted image of "value-free" science.

In the milder forms of naturalism, we see that their reconciliation of descriptive and normative inquiry begins by easing the strictures against the old philosophical sin known as "psychologism," allowing a fully normative naturalism to emerge. Traditional analytic epistemology distanced itself from empirical inquiry, seeing its job as a search for the logical properties and relationships of propositions that justify some true beliefs. This traditional epistemology gives an apsychological account of what it means to be justified in believing a proposition in which the required logical relationships between beliefs can be specified without making use of psychological methods, results, or concepts (Kitcher 1992, 56–57; Kornblith 1994, 133). Naturalized epistemology rejects this rejection of psychology, seeing the empirical study of how we actually *do* form our beliefs as relevant to the normative question of how we *ought to* form our beliefs. This naturalistic approach does not seek to collapse epistemology into psychology but instead makes use of the results of psychology and cognitive science to help answer questions about which techniques for belief formation work reliably.

A slightly different way of seeing empirical inquiry as relevant to normative epistemological concerns can be found in Larry Laudan's notion of a "normative naturalism," which makes use of the history of science to seek testable methodological norms. In Laudan's view, a methodology is a set of rules or maxims to be understood as hypothetical imperatives. These imperatives say: "If you want to achieve A, then do B." Methodological norms are thus instrumental maxims that rest on (among other things) claims about the world and are to be tested and chosen in the same way we test or choose empirical theories. If there is good evidence that following a rule will help us achieve our aims, then we have a good reason for endorsing that rule. The historical record can thus serve as a source of evidence for testing these hypotheses (Laudan 1987, 24–27; see also Giere 1999, 72–75). Of course, the "history" appealed to in examining knowledge borrowed from chaos theory will be extremely recent history.

In characterizing normative naturalism, Brown answers the question about the relationship between descriptive and normative when he says, "Norms need not be trans-empirical. Rather, norms, in the forms of both ends for science and methodological imperatives, are introduced and evaluated in the same ways as theoretical hypotheses, experimental designs, new mathematics, and other features of the so-called content of science. People propose them and try them out. If a proposal looks promising it is pursued, and if its promise vanishes it is modified or dropped" (Brown 1988, 69). There is no problem of trying to deduce "ought" from "is," because no one is attempting a deduction at all: norms are hypotheses that are proposed and tried and tested. Scientists do this all the time, so why not philosophers (75)? Ultimately, the normative naturalism of Laudan, Kitcher, Brown, and Giere lives up to its title, and Laudan is correct when he encourages us to see "normative and descriptive concerns interlaced in virtually every form of human inquiry. Neither is eliminable or reducible to its counterpart" (1990, 56). Normative naturalism represents a truly pluralistic form of interdisciplinary epistemology, and it speaks to the interaction of facts and values that I explore further in chapter 7.

Defenses and Reformations

Disciplinary pluralism rejects both imperialism and isolationism in interdisciplinary relations. As we have seen, normative naturalism rejects imperialism by seeking not to absorb or eliminate other forms of inquiry but to make use of them when they are helpful. We have also seen how disciplinary pluralism overcomes the isolationism of attempts to draw a sharp line between the context of discovery and the context of justification. Another form of isolationism is manifested in the work of philosophers who say that epistemology ought to deliver norms that are discoverable a priori. Such a claim means that epistemology is a form of inquiry independent of experience, and thus it is a discipline that is intrinsically separate from empirical disciplines, with nothing to learn from them. So how can a pluralistic and naturalistic account of methodology and epistemology answer the charge that it fails to do justice to a priori knowledge and the independence of epistemological norms from the facts of experience?

One way is simply to deny that there is a priori knowledge. Kitcher seems to take this approach when he declares that "virtually nothing is knowable *a priori*, and, in particular, no epistemological principle is knowable *a priori*" (1992, 76). But there is another way to deal with the challenge, which is to reconceive of a priori knowledge. Brown suggests that a priori knowledge comes from analytic truths about our conceptual structures and hence is independent of experience in some sense. But our concepts are never strictly forced upon us by experience, and we may make a pragmatic decision to change our concepts because of our experience, so the a priori is changeable. Epistemic norms could be based on conceptual truths, and thus established a priori (perhaps by simply reflecting on our concepts), but such norms are still subject to empirical control and reevaluation at some level (Brown 1988, 55–64).[12] Thus, there can be conceptual inquiry that is relatively immune to the deliverances of experience and yet not completely separate from empirical inquiry. Indeed, as Laudan points out, naturalistic philosophy need not be narrowly empiricist because science is not. Conceptual issues matter for science, and philosophy can use the same methods as science (1990, 50).

As a form of naturalism, disciplinary pluralism does not claim a priori status for its deliverances, but neither does it rule out a reconceived form of a priori knowledge. Naturalism need not imply that logic and mathematics are empirical in exactly the same way that the rest of knowledge is, and that philosophy is merely a branch of science—such inferences would be another form of imperialism. A disciplinary pluralist will be perfectly happy to accept that some pieces of knowledge function more like modes of representation than others and that philosophy has a distinctive task of arguing about what is to count as a fruitful way to adjust to experience. So, without completely rejecting the possibility of a priori knowledge, we can still recognize that at least some epistemological and methodological normative questions cannot be investigated independent of experience.[13] Although there may be a place for a reconceived a priori, investigation into scientific knowledge and its uses should not be done in total isolation. Traditional epistemologists will not be satisfied with this account of the a priori, or indeed with naturalistic approaches in general, because they see them as reneging on the fundamental questions of epistemology that treat the foundations of human knowledge and their justification in a uniquely philosophical sense. There may be no adequate reply to this concern, except to point out that normative naturalists are unanimous in letting go of such justificatory projects as the Cartesian need to answer the skeptic (Kornblith 1994, 12; Brown 1988, 75; Giere 1999, 76; Kitcher 1992, 63).

Before we take up normative naturalism in our study of borrowed knowledge, two important modifications must be made: it needs to be socialized, and it needs to be made more fully pluralistic. Naturalism still has a strongly individualistic bias. We see this when Kornblith and others deal almost exclusively with the psychological investigation of individual knowers in discussing empirical inquiry into knowledge making (Kornblith 1994, 4). Steve Downes argues forcefully against what he calls cognitive individualism, "the thesis that a sufficient explanation for all cognitive activity will be provided by an account of autonomous individual cognitive agents" (1993, 452). Scientific cognition, the activity that leads to accepted, published claims, includes more than the processes that go on in individuals' minds—it also includes material representations, experimentation, the creation and testing of hypotheses, peer review of publications, and the like. Attention to these processes will widen the scope of naturalistic philosophy of science, although some naturalistic philosophers of science may see the social as irrelevant to the production of knowledge or, worse, as a corrupting influence. Downes cites Rouse (1987), Hull (1988), and Longino (1990) as philosophers who have socialized science without simply reducing its content to social interests, and urges that we not be satisfied with descriptions of science solely in terms of what goes on in individual scientists' heads (1993, 467). In a pluralist spirit, we can admit that for some purposes, on some occasions, we may in fact be satisfied with a purely individualistic approach. But in the following chapters, I endeavor to pay due attention to the social aspects of the phenomenon of borrowing.

Second, we should heed David Stump's call for a truly pluralistic naturalism in the examination of science. Naturalizers sometimes pick only one discipline as a source for epistemological insight: cognitive psychology, history, evolutionary biology, or sociology. But science does not have just one method, so why should the study of science (Stump 1992, 457–58)? In what follows I endeavor to exemplify a normative naturalism that fully embraces disciplinary pluralism, making use of a wide variety of approaches, both scientific and nonscientific, where they are appropriate. Rhetoric, sociology, and history help make sense of the social function of borrowed knowledge. Cognitive science, even with its individualistic tendencies, is useful for understanding metaphor. Formal methods will not make much of an appearance, but they certainly have their place (e.g., see Kittay 1987, 172–74). There may even be some conceptual analysis along the lines of a priori reflection. Nonetheless, there is a limit to the number of disciplines one person can make use of. The absence of an anthropological perspective in this study, for instance, does not indicate a failure of pluralism but rather a limitation on time, space, and training. Disciplinary pluralism is ultimately a feature of a community of inquirers—a community that this work seeks to inaugurate.

3 The Rhetorical Functions of Borrowing and the Uses of Disciplinary Prestige

A pluralistic methodology for science studies faces the challenge of bringing together disciplines that may seem unlikely to cooperate. Consider for example the difficulties involved in accommodating both philosophy and rhetoric, two fields of inquiry that have generally held each other at a distance for millennia. Yet, in addition to a broadly philosophical orientation, a rhetorical approach helps inform a multidisciplinary examination of borrowing by drawing our attention to the actual functions it serves in academic inquiry. After all, researchers do not produce knowledge in a vacuum; they need to get their work published, convince their colleagues to take them seriously, and respond to criticism, all of which are tactics treated by rhetoric. Borrowed knowledge helps with all these tasks of persuasion, and rhetorical analysis can illuminate how borrowing serves these functions.

In what follows, I first sketch what I consider to be the rhetorical perspective, especially as it relates to scientific inquiry, and treat one major source of conflict between rhetoric and philosophy: the question of normative evaluation. Next, I clarify how the rhetorical functions of borrowing from chaos theory trade on the newness of this field and the disciplinary prestige of the natural sciences. Finally, I conclude with some reflections on the problems that can arise when researchers borrow from disciplines as prestigious as the natural sciences. Throughout, my concern is with the persuasive functions served by borrowed knowledge. Persuasion is not here meant to require emotional manipulation, deception, coercion, or irrational conversion. At least since Kuhn, science studies have recognized the role of persuasion in theory change. Indeed, Richard Boyd says that "Kuhn's work has made it clear that the establishment of a fundamentally new theoretical perspective is a matter of persuasion, recruitment, and indoctrination" (1993, 488). The use of persuasion need not mean a lack of concern for making reliable knowledge; it can be either misleading or appropriate. Among its appropriate uses are the legitimizing functions of justifying new methods and making new ideas acceptable, as well as the functions of promoting one's ideas and simply getting them published.

The Rhetorical Approach

What does it mean to take a rhetorical approach? Although traditional rhetoric may have begun as a study of how to make effective orations in public gatherings, much recent work in the field has conceived of rhetoric as a lens for critical analysis. Chaim Perelman has been a key figure in this updating of the rhetorical tradition, redirecting it toward investigation of the use of discursive techniques that secure agreement (Perelman and Olbrechts-Tyteca 1969, 4). Let us consider each of the three parts of this characterization: use, discourse, and agreement. First, as an example of the attention paid to use, Alan Gross points out that rhetorical analysis looks at Darwin's *Origin of Species* as practical knowledge, as a means of getting something done: "the vehicle by means of which Darwin attempted to persuade his fellow biologists to reconstitute their field, to alter their actions or their dispositions to act" (Gross 1990, 5). So the rhetorical approach looks at the instances of borrowing and asks the following: What are the borrowers trying to accomplish? How are they using knowledge from the physical sciences to get something done?

Next, rhetorical analysis is crucially concerned with discourse, as when Jeanne Fahnestock asks, "How do the structures or options available in a language lead us into certain prepared lines of thought or argument?" (1999, vii; see also xi). This attention to language encourages us to look at figures of speech, especially metaphors as in chapters 5 and 6, and also to consider words, phrases, and entire arguments, as discursive resources. As a text-based inquiry, rhetoric may investigate notebooks, letters, manuscripts, and occasionally diagrams and illustrations. It does not typically encompass the investigation of artifacts, networks of mentorship, or the flow of money, for example. So the rhetorical perspective, concerned with the use of language, leads us to see chaos theory as a stockpile of discursive resources that can be mobilized for certain goals.

Third, rhetoric looks at techniques, specifically arguments, that are used to secure agreement among people. Here an "argument" does not mean what it does in formal logic, that is, a set of propositions and their strictly analytic connections and entailments. Much traditional philosophical training considers arguments abstracted from their context, whereas a rhetorical approach represents a complementary enterprise that cannot be engaged in simultaneously, as discussed in the previous chapter. Indeed, rhetoric cannot make sense of the very idea of context-free evaluation because it examines utterances by individuals in specific situations who are seeking to win the agreement of actual listeners. This aspect of the recent rhetorical approach takes its cue from Stephen Toulmin's *The Uses of Argument*, which looks at what he calls "the practical assessment of arguments," including their structure, merits, and how to criticize them (1958, 2). In this approach, each argument necessarily takes place within an argument field—the actual social and communicative context of argumentation (14; also Willard 1983, 10).

Some of the central concepts used in rhetorical analysis have been identified as style (for example, questions about the choice of words), arrangement (for example, questions about the order in which ideas are presented), and invention (questions about the ways conceptual resources are used to generate new ideas) (Gross 1990, 69). Other elements of the rhetorical approach include the concepts of genre and the store of topics, but these are less important for our purposes here. Two crucial rhetorical concepts that play a significant role in what is to come are the notions of *audience*—those who provide the required context for persuasion—and *authority*—the social legitimacy that enables one to be taken seriously (Perelman and Olbrechts-Tyteca 1969, 19; Gross 1990, 13). Although in this volume I do not use a lot of the terminology or techniques of classical rhetorical analysis, I emulate the branch of rhetorical studies known as the "rhetoric of inquiry." The rhetoric of inquiry takes the concepts and techniques of rhetorical study and uses them as a framework for examining contemporary scholarship (as opposed to, say political persuasion or advertising). When that scholarship is scientific work, the approach is called the "rhetoric of science."

The rhetoric of inquiry approach sees knowledge as the outcome of social, communicative processes of persuasion within which logical rigor, epistemological standards, methodological rules, evidential salience, and relevance are negotiated schemes for legitimating beliefs, and thus "essentially rhetorical" (Gross 1990, 4; see also Simons 1990; Scott 1967). Rather than seeing scholarly controversies as settled simply by the use of standard methods or the appeal to straightforward facts, the rhetoric of inquiry examines the written and spoken discourse that actually serves to settle the matter. In keeping with the earlier discussion of methodological pluralism, we should be wary of the disciplinary imperialism that seems to lurk in the claim that knowledge making is "essentially" a rhetorical process. Such a claim might be taken to mean that the production of knowledge is "merely" or "entirely" a matter of persuasion. The more modest claim would be that knowledge production "always also" involves persuasion. So the modest position holds that rhetoric provides a useful perspective because the making of knowledge uses words in a social process of reaching agreement by persuasion. Although the rhetorical approach is appropriate for answering certain questions, it need not claim to be wholly sufficient. The need for judgment in inquiry, the symbolic mediation of facts and reasoning, the role of communities, values, and language: all of these make a place for a rhetorical investigation of inquiry but do not support the imperialist claim that there is no place for other approaches (Simons 1990, 2).[1]

There is an important difference between much of the scholarship in the rhetoric of inquiry, and specifically in the rhetoric of science, and what I intend to be doing. Much of the rhetoric of science examines the persuasive practices that scientists use on each other to establish the justification of their claims. Thus, they are interested in showing that even in the "hard" sciences there are elements of such "soft" rhetorical elements as employment of figurative language, strategies for establishing authority, and appeals to shared values. I am more interested in the persuasive rhetorical power that science has for helping people in other disciplines persuade their colleagues of their results.

And although there are metaphors, authority, values, and so on in scientific inquiry itself, they do not delegitimize or exhaust science. As literary scholar Ira Livingston cautions, "the notorious unscientificity of metaphor may enable the 'gotcha' effect of showing the absolute reliance of science upon metaphor" (1997, viii), but such exercises rest on misunderstandings of both science and metaphor. Recall my distinction between the moderate claim that knowledge production is always at least partially a matter of persuasion and the extreme claim that knowledge production is never anything except a matter of persuasion. Some may claim that science is merely a powerful persuasive technique, or just the superstructural expression of social values, or simply a collectively agreed-upon metaphorical construction of the world, yet these meta-scientific monisms are exaggerated and imperialistic. They point to a need for disciplinary pluralism. As the rhetoric-of-inquiry scholar John Lyne counsels us, we need to get beyond just looking for the presence or absence of rhetoric, and we should recognize that rhetoric alone cannot account for everything that inquiry does (1990, 55). The converse of this idea, of course, is that if authority, metaphor, and values can be found in science itself, then the fact that they are present in economics, law, and literary studies ought not delegitimize these fields of inquiry either. If science itself does not live up to the ideals of a cartoon version of positivism, then we need not denigrate other fields for their failures to do so. This statement reinforces the notion that economics, law, and literary studies are also rational enterprises that make knowledge.[2]

Before considering a possible objection to the rhetorical approach, pluralist scruples demand an acknowledgment that other approaches could prove equally illuminating. Two other perspectives that show great promise are the sociology of knowledge and—perhaps more surprisingly—marketing theory. Sociology examines systems of power and can illuminate the strategic decisions made by individuals who must navigate these social systems. A sociological analysis might begin by contacting a large sample of researchers who have made use of chaos theory in journal articles outside the natural sciences. A survey might ask the following: When and how did you become aware of chaos theory? When did you decide to make use of it? What benefits or drawbacks did you anticipate in doing so, and what benefits or drawbacks have you since experienced? How has your work been received? A statistical analysis of the responses to such a survey could look for patterns in the data. Perhaps, as one sociological study has indicated, the use of chaos theory peaked at a certain point in time (Weingart and Maasen 1997). Perhaps there are interesting differences in the amount of borrowing by practitioners in different fields or at different stages of their careers. In-depth qualitative interviews could complement such quantitative studies, more fully exploring any patterns revealed by the initial data and plumbing the responses for stories, concerns, or attitudes not initially revealed. But a pluralist approach requires one to make practical decisions about the scope of multidisciplinary endeavors; limitations of time, money, and training require me to leave sociological investigations to others. And that is probably for the best, because any scholar receiving a survey from me would be able to discover with a quick Web search that I have already expressed some strong opinions in evaluating certain instances of borrowed knowledge.

Sociology is, of course, only one of several disciplines that could be employed to further understand the practice of borrowing knowledge. The theory (and indeed the practice) of marketing has already been applied to the academic world. It is all too common to hear university administrators speak of the "higher-education marketplace" and of the need to position, promote, and brand a school's product to the "consumer." Here the consumer is often conceived of as the student or parent paying for a college education, but sometimes the institution's graduates are themselves cast as the products who need to be tailored to fit the needs of their ultimate consumers—the corporations who will employ them. But with respect to borrowed knowledge, the relevant market is the "marketplace of ideas"—specifically, ideas published in academic journals. A marketing perspective would examine the production of journal articles and scholarly books as a competition for such limited resources as publications, grants, and spots on conference programs. The study of marketing includes product design, production, promotion, and distribution, which could be applied to their analogues in the world of academic publishing. The language of marketing could provide a useful source of metaphors for academic pursuits, and the construction of an extended analogy between academia and marketing is a promising project for understanding the actual processes of knowledge production. But such a detailed analogy would initially take us far from the details of borrowed knowledge and from my interest in evaluative questions, so I leave it aside for now.

Rhetoric and the Normative: A Problem of Relativism?

A potential challenge to the use of a rhetorical approach arises from its descriptive character: by limiting itself to a description of the resources that in fact achieve assent, rhetoric seems to imply that whatever works, works. Such an approach raises concerns about a relativism that would make it impossible to criticize some instances of borrowing as better or worse than others, similar to the challenge faced by normative naturalism in the previous chapter. One purpose of this project is to evaluate these borrowings critically, so readers would have cause for serious concern if a rhetorical approach hamstrung any attempt to reach this goal. In this section, I begin by setting out the problem as clearly as possible, clarifying how this problem arises specifically within the rhetorical approach. Second, I outline a defense of the rhetorical approach against the claim that it rules out the possibility of evaluation and criticism. And finally, I discuss the kind of normative force that this type of investigation can yield.

The problem about rhetoric and relativism has its roots in the distinction between the descriptive and the normative, which in turn relies on a supposedly unbridgeable gulf between facts and values. In chapter 7 I examine the possibility of building argumentative bridges across this gulf. For now, it is enough to recall the discussion of naturalism in the previous chapter and keep in mind the compatibility of descriptive investigation and normative criticism. Recall that for Aristotle, rhetoric was a craft, a *techne*, and hence always normative because it treats the questions of better and worse arguments.[3] Why, then, does the problem of relativism arise for the rhetorical approach? To answer this question, I look largely at the work of Chaim Perelman, who is central to the revival of rhetorical theory and to contemporary work in the rhetoric of inquiry. Perelman and Olbrechts-Tyteca cite John Stuart Mill as making the distinction that evidence is not "that which the mind does or must yield to, but that which it *ought to* yield to, namely, that, by yielding to which, its belief is kept conformable to fact" (1969, 3). Here we see the crucial distinction for epistemology between a description of what does bring agreement and an evaluation of what ought to bring agreement. Perelman reconfigures the distinction. For Perelman, "an efficacious argument is one which succeeds in increasing this intensity of adherence among those who hear it in such a way as to set in motion the intended action" or a disposition to do so (Perelman and Olbrechts-Tyteca 1969, 45). If we are only interested in what is effective, what actually works to obtain agreement, it seems that we are being relativist because we refuse to pass evaluative judgment on the arguments used to achieve that agreement.

For Perelman, such judgment is beside the point because his revived rhetorical analysis stays at the level of the descriptive. Similarly, McCloskey claims that rhetorical criticism of economics "is not a way of passing judgment on economics. It is a way of showing how it accomplishes its results" (1985, xix). And Weingart and Maasen pursue an empirical investigation of the uses of chaos theory that eschews passing judgment on instances of borrowing. Using techniques of discourse analysis and the sociology of knowledge, they seek to "look for the *variety* of applications" rather than "*appropriate* applications" of chaos theory, and assert that "on this view, nugatory uses cannot be found" (1997, 473). These empirical approaches look at how inquiry happens, not at how it ought to happen.

One of the most radical expressions of this view comes from Charles Willard, who explicitly rejects the difference between "genuine" and "conventional" knowledge and encourages us simply to seek an "empirically adequate sociology of knowledge" (1983, 4). Willard suggests that the study of knowledge claims ought to pursue descriptive rather than normative goals, and he is hostile to evaluation and normative epistemology, claiming that those who still want to do it are either running scared from the boogeyman of relativism or being "held hostage to social preferences and favored practices" (3–5, 16). In fact, his stated goal of simply describing and explicating actors' evaluative practices takes on an air of supposedly scientific objectivity (17). For Willard, the two key questions are how do people get their ideas to pass muster by adopting and employing the standards of an argument field and how do people come to trust the authority of certain people and standards (22)? Willard is especially instructive because he is a strong proponent of the descriptive approach, yet we will see that even for Willard the rhetorical approach need not rule out evaluative criticism.

Another potential source of relativism in the rhetorical approach is the claim that, as Toulmin puts it, there are no field-invariant standards for evaluating the strength of arguments (1958, 38). Willard pushes this point even harder when he claims that each argument field is autonomous (1983, 11). If the strength of an argument depends on its context and its audience, on the shared assumptions and practices of a community of inquirers, then this dependency seems to undercut the possibility of making normative claims of an argument being good or bad: it is only good or bad for some specific audience at some specific time for some specific purposes. Among historians, this sort of relativism may arise from a concern to avoid applying current standards to past practices. Such a concern can give rise to an extreme historicism, which holds that each event can only be understood within its unique historical context. Among sociologists of knowledge, relativism may be an explicit methodological injunction. In the sociology of science, this injunction is often associated with the "symmetry principle" that dictates that sociological explanations for the acceptance of a piece of scientific knowledge should be rigorously impartial with respect to truth and rationality (Bloor 1976). In these three empirical disciplines, radical contextualization undercuts the ability to make normative judgments.

However, philosophical pronouncements by scholars in empirical disciplines need not match up with the actual work that they do. After all, claiming that you are not a relativist does not mean that your position does not entail relativism. And conversely, claiming to be a relativist or trumpeting your rigorous avoidance of passing judgment does not always square with actual research practice. We have to look and see. Using a rhetorical (or sociological) perspective does not commit one to a relativist epistemology—or a constructivist metaphysics, for that matter. Using an empirical approach to examine the phenomenon of borrowed knowledge does not require one to swallow the imperialist slogans that rhetorical theorists and sociologists of knowledge sometimes bandy about.

Consider Willard. His relativism serves to motivate attention to actual communicative practice (see esp. 1983, 8–9). Within that practice he is concerned with making judgments. He talks of "studying the ways fields get their business done and inquiring into ways of improving these methods" (233). And he insists that "we still want to know how arguments yield knowledge. We want to know a good argument when we see one; and we want to know why it is true" (279). But he encourages us to look to well-argued consensus, not formal logic, as a source for evaluation. Similarly, Weingart and Maasen consider that a metaphorical use of chaos theory can be "a very superficial usurpation" or suffer from a self-contradiction that renders it less promising (1997, 496, 502).

Another rhetorical theorist, Herbert Simons, identifies a reconstructive rhetorical approach that asks what is good judgment, what are good reasons, and what are appropriate forms of argument? And for Simons, as for Willard, the rhetoric of rhetorical approaches serves to carve out a space for investigations of knowledge that are not amenable to formal approach. Reconstructive rhetoric moves beyond merely debunking to consider "how one ought to argue and use language in situations and on issues for which there can be no proof in the strict sense of that term" (Simmons 1990, 6).

The recognition that standards of judgment escape formalization does not necessitate a relativism in which judgment is impossible. Similarly, the recognition of the contextual nature of argumentation need not yield a relativism impotent to make evaluations, because it is sometimes possible to compare across contexts. Granting, as does Douglas Walton, that the presence of multiple normative frameworks for different dialogical contexts creates a kind of relativism does not render critical evaluation impossible (1997, 21–22). The political scientist John S. Nelson speaks of "immanent grounds of criticism," which "can in principle and do in practice combine with comparisons across inquiries to respect contexts without idolizing them" (1990, 264). Nelson cites McCloskey as an example of this comparative approach, and McCloskey insists that the rhetorical approach does not entail relativism or put an end to rigor, precision, and serious thought (1985, 35). Again, to insist that one's position does not entail relativism is not a sufficient argument that one is not in fact a relativist. But the detailed work by McCloskey in, for instance, critiquing faulty statistical reasoning makes it clear that attention to patterns of argumentation need not make one accept whatever works.

Sometimes rhetoric (and occasionally the sociology of science) invokes relativism to make room for its perspective on knowledge making. In part, the insistence that standards of argumentation are relative to an argument field frees the analysis of argumentation from an exclusively formal approach that aspires to universal, context-independent standards of appraisal. This strategic or methodological "relativism" that rejects the search for context-free standards should not be confused with the relativism that would deny the existence of any standards whatsoever. Furthermore, rhetorical theorists may need to invoke relativist slogans in response to the metaphysical realist slogans invoked by some philosophers of science (and some scientists) who sometimes sound as if the only explanation necessary for why scientists arrived at a particular result is that it was the truth. Relativism, symmetry, and even rhetorical imperialism are sometimes useful ways to insist on a space for alternative approaches (see, e.g., Willard 1983, 8–9). The insights and methods associated with these approaches can be used without acceptance of the accompanying slogans (see Fine 1996; Longino 2002).

Rather than engaging in aprioristic epistemology (seeking to discover from first principles what counts as a genuine justification) or purely descriptive rhetorical study (looking for those arguments that actually persuade), I investigate borrowing by looking at the arguments and techniques people actually use and criticize them because some are better than others—a kind of bottom-up normativity. This approach is pragmatic, provisional, and instrumental, setting out the benefits of certain strategies for achieving certain goals and the dangers of other approaches. Advice such as this recalls the instrumental conception of norms discussed in the previous chapter's treatment of normative naturalism.

So, instead of seeking to solve the problem of relativism, we might rather engage in the hard work of criticism, proceeding on a piecemeal basis according to detailed investigations. Instead of setting out at the beginning by dictating "Here is what knowledge is, and here is how inquiry must proceed in order to meet these preestablished criteria," let us instead examine what people are doing. One scholar wants to accomplish a particular aim in law, and another has a particular goal in literary interpretation. Well, do they accomplish their goals? And are their goals good ones to have? Pluralism does not presume that there is just one answer to the question of how to make good knowledge, and so normative standards and evaluation emerge from actual investigation. It is to this investigation that I now turn.

The Rhetorical Uses of Borrowed Knowledge and the Power of the New

Rhetorical inquiry can investigate persuasive, inventional, and expository purposes. That is, for example, an author's borrowing of knowledge from the physical sciences can help persuade others to accept or consider new ideas—serving a persuasive purpose—but it can also help researchers to come up with new ideas in the first place—serving as an inventional resource. Borrowed concepts can also serve an expository function by helping to explain or illustrate difficult ideas in the target discipline. This rough division provides a preliminary way to structure an investigation of the purposes served by borrowed knowledge.[4] I pass over the expository use of borrowed knowledge, for it is rarely the case that notions in the social sciences and humanities can be more clearly explained by analogy with physics. As several writers have noted, the only time it is helpful to clarify a discussion of law or literature by constructing a parallel illustration from nonlinear dynamics is when one is addressing an audience of mathematicians or physicists (Niman 1994, 362; Sokal and Bricmont 1998, 10, 133; Matheson and Kirchhoff 1997, 41). Yet metaphors can serve as more than just illustrations or pedagogical aids. In this chapter and the next I look at persuasion, whereas in the following two chapters on metaphor I look more at invention. The first cluster of persuasive functions served by borrowing from chaos theory derives from the simple fact that chaos is such a new field.

The Power of the New

In examining many borrowings from chaos theory, it is striking how often terms from physics get tossed in merely for flavor. Gross and Levitt decried the use of chaos theory as a source of decoration in the humanities with some good reason (1994, 95). We can easily find examples of this embellishment in some works of literary interpretation, as when Harriet Hawkins gives a perfectly fine discussion of Shakespeare that stands independent of anything chaotic but then throws in a few technical-sounding terms borrowed from recent work in nonlinear dynamics (e.g., 1995, 137, 141). What persuasive purpose is served when a scholar simply replaces "repeat" with "iterate" or "similarity" with "self-similarity"? What is the point of inserting words like "iteration" into a recapitulation of Morrison's *The Bluest Eye* or a close reading of Cormac McCarthy, or of merely observing "butterfly effects" in various novels (Slethaug 2000)? Why does Ira Livingston pepper his perfectly good exploration of romanticism and postmodernism with terms such as "fractal" and "strange attractor," employed without explanation?[5]

In a survey of the rhetorical purposes served by chaos theory, we can start with the undeniable appeal of anything that seems new and different or fashionable and trendy. The pattern of argumentation that rhetorical analysts call "the topic of quality" proves useful in examining the persuasive power of borrowing from a cutting-edge science like the chaos theory of the1990s. A "topic" is a readily available argument scheme—in this case, the scheme that characterizes whatever is new as being better. Novelist John Barth alludes to this topic of quality when he admits that one of the reasons to use chaos theory is that "one writes a contemporary novel by writing it in a contemporary way" (1995, 341). As in literature, so in literary theory or law or economics—chaos theory serves as a sign of the new. And being new can help you get past one of the first hurdles of academic knowledge production by getting you published, read, and taken seriously.

Audibility

Borrowed knowledge can serve the important rhetorical function of helping to meet the requirement I call "audibility"—the need for an audience. Whereas "credibility" calls to mind a speaker's desire to be believed, "audibility" refers to the need to be heard (or read, or clicked on) in the first place. Here we see a part of the process of making knowledge that comes before justification and constitutes the first step in the context of pursuit: simply getting attention. As Perelman and Olbrechts-Tyteca point out, "it is not enough for a man to speak or write; he must also be listened to or read. It is no mean thing to have a person's attention, to have a wide audience, to be allowed to speak under certain circumstances, in certain gatherings, in certain circles" (1969, 17). Although this need can be met by specialized journals, many submissions to journals are rejected, and many published articles go unread. In the flood of information in the academic marketplace, potential audience members must screen the many candidates seeking their time and attention. This screening should not be considered an entirely irrational activity; neither is it irrational to compete for attention. Although academic writing is not entirely a function of marketing, we cannot ignore the fact that attention is a scarce resource. The rhetorical perspective, which insists

on considering the audience for each written text, allows us to see that
the audience for chaos theory (or any scientific knowledge) is more complicated than just "scientists" and "the public" (Lyne 1990, 51). Besides
these two groups, the audience includes potential readers of law review
articles, books on literary interpretation, and journals of economics and
business. Academic authors may not like to admit that we must sometimes resort to marketing techniques, but the power of the new can be
very helpful. Frank Miele bemoaned the fact that that for humanists and
social scientists, "uttering the words 'chaos' and 'complexity,' especially
in the right tone of voice and at the right time, can get your way paid to
an academic conference in some fashionable part of the world" (2000,
60). Yet physicist David Ruelle felt it necessary to end an article on chaos
theory in *Physics Today* with a plea to stem the flood of publications exploring chaotic behavior, urging that "just writing another research paper is not considered a good reason" to study dynamical systems (1994,
29). His urgency provides evidence that there was a problem even in the
natural sciences of people hopping aboard the chaos theory bandwagon
in order to get insignificant work published.

Chaos theory is new, or at least it was new not too long ago. But
invoking any science in a radically unusual context is a tactic that also
plays on the appeal of the new. Thus, interdisciplinarity itself can serve
to get the attention of one's chosen audience and also to broaden one's
potential audience. John Dupré has identified the lure of generality as
one of the factors that help to illuminate the attractions of borrowed
knowledge. He suggests that cross-disciplinary ideas that promise great
breadth of application may "seem more exciting" and have "more
chance of making the bestseller lists" than those with less exaggerated
scope (Dupré 2001, 81). For a postmodern literary/cultural theorist like
Ira Livingston, references to chaos theory serve both to evoke the up-to-
the-minute thrill of cutting-edge scientific developments and to perform
the writer's identity as a bold transgressor of disciplinary boundaries.
Livingston's book *Arrow of Chaos: Romanticism and Postmodernity*
(1997) appeared in a series bearing the name Theory Out of Bounds "un-
contained by the disciplines, insubordinate . . . inventing, excessively, in
the between . . . Practices of Resistance; Processes of Hybridization."
Crossing boundaries indeed presents many marketing advantages. Perelman has pointed out that because an audience is composed of many
different elements, a person skilled at argument must use a variety of
arguments, and an argument must be adapted to what is appropriate for
its audience (Perelman and Olbrechts-Tyteca 1969, 22, 25). I contend
that what Perelman says of one's actual audience also holds true for

one's potential audience: a varied appeal in terms of multiple disciplines can widen the possible market and opportunities for attention. So even if borrowing chaos theory is not necessary to make one's point and does not actually strengthen the case, it can bring one's work to a broader readership. Borrowing allows for an appeal not just in terms of a particular discipline, but in terms of another radically different discipline as well. It certainly does not hurt that the other discipline being appealed to is one with tremendous prestige. The role of the prestige of science is treated later, after first discussing the power of the new for securing legitimacy.

Legitimacy

The new may often be worthy of attention, but rarely is an idea worthy of acceptance merely because it is new. Yet sometimes those borrowing from recent scientific research rely on the topic of quality to confer on their argument not just audibility but credibility. Matheson and Kirchhoff identify this tendency in their critique of work on chaos theory in literature: "to label a theory as a new paradigm is to empower it; it is to say that the theory represents an earth-shaking innovation with ramifications that extend far beyond its intended field—for instance, it helps to legitimate claims about the general cultural relevance of that theory" (1997, 31). But they go on to say that "a denial of chaos's status as a new paradigm provides *prima facie* reason to doubt its cultural significance," and here they may have overreached. Plenty of new scientific discoveries that do not rise to the status of new paradigms may have significant cultural implications.[6] In the quest for legitimacy, the new and unfamiliar can help to intimidate potential critics, or at least keep them off balance. For example, my use of the image of cross-training illustrates the intimidating power of an unfamiliar metaphor: because many philosophers will be caught off guard by an illustration from the world of sports, they may be less able to conjure up quick counterexamples.

The power of the new may sometimes be invoked as a condemnation of the old. Constitutional scholar Laurence Tribe seems to fall into this pattern in his essay "The Curvature of Constitutional Space: What Lawyers Can Learn from Modern Physics." Tribe does not appeal to chaos theory in his argument for a new paradigm of legal thought, but he does suggest that some recent constitutional rulings "might well reflect a partial throwback to a more primitive paradigm" (1989, 26). Regardless of our individual views about scientific progress, we can recognize here that the term "primitive" serves to cast aspersions on older ways of thought. Royce de R. Barondes takes Tribe and others to task for their tendency to use "Newtonian" as a term of abuse, claiming that such formulations lead "to an enticingly easy conclusion that any analysis labeled 'post-Newtonian' is preferable" (1995, 176). Certainly an idea is no more likely to be true simply because it has been fashioned by analogy with a more recent piece of physics. But as we see in the next chapter, the power of the new can be invoked to help motivate methodological change without making illegitimate claims to epistemic merit.

Curiously, many literary scholars forge links with chaos theory that appeal not to its newness but to its similarities with old (or even ancient) knowledge. For example, Thomas Jackson Rice states that "In *Ulysses* James Joyce uncannily anticipates the perspective of the new scientists of chaos" (1997, 83), and others have found such anticipations or "prefigurations" in romantic aesthetics, Nietzschean philosophy, postmodern novels, or the culture in general.[7] What is the rhetorical function of such claims of anticipation? We can divine a clue as to its purpose from Thomas Weissert's pointed summary of a pronouncement by Michel Serres: "great literature often discovers scientific truth long before the scientists get around to it" (Weissert 1991, 223). Impatience with science, which takes so long to realize what literature already knows, becomes explicit in the work of Harriet Hawkins. She claims not only that chaos theory enables us to gain new insights into classic texts, but that by finding order governing apparent chaos this new science incidentally allows "people familiar with mythic literature to say 'we knew it all along'" (1995, 7). Notice that here the new knowledge produced by science is taken as confirmation and vindication of traditional insights. Looking ahead to the discussion of disciplinary prestige, note that the intellectual authority of the sciences is borrowed here to support an existing claim of knowledge. Furthermore, the prestige of science allows scholars of the classics who may feel neglected or underappreciated to claim priority in making an important discovery. It makes their field seem relevant and necessary when they have come to feel under attack for being outdated.[8]

The Rhetorical Uses of Borrowed Knowledge: Disciplinary Prestige

The persuasive use of borrowed knowledge from the natural sciences seems to rest on the intellectual and cultural authority of these fields. Does borrowing knowledge thus constitute a fallacious appeal to authority? Rhetoric scholar Douglas Walton has argued that although the "argument from authority" appears in many logic textbooks as a fallacy, things are not so simple. Appeals to authority can indeed be fallacious, especially when they are used to shut down further questioning, but in reasoned argumentation, expertise can generate a presumption or shift the burden of proof while remaining open to critical questioning. If the questions raised by the use of a particular authority are answered acceptably, then the claim is provisionally acceptable. Thus, appeals to authority represent a legitimate and defensible form of reasoning (Walton 1997, 31, 228).

Indeed, appeals to authority show up in science itself. For example, the author of an article in a scientific journal often invokes past results to argue for the importance of current work or cites past practice to argue for the credibility of current methods. Applicants for grants will demonstrate that their previous work has appeared in respected journals, that they have received other grants, and that they are part of an established research institution with an ongoing research program (Gross 1990, 13). In general, the division of cognitive labor requires scientists to appeal to the authority of their colleagues, because modern cooperative scientific investigations are often far too complicated for any one person to master all the required areas of expertise (Hardwig 1985; see also Kitcher 1990).

Yet argument from authority, or *argumentum ad verecundiam*, is not really what is going on in borrowing. Instead of trying to establish a proposition based on the personal credibility of experts in the natural sciences, the author may use borrowed knowledge to appeal to the intellectual authority of the natural sciences themselves.[9] I call this practice *invoking disciplinary prestige* in recognition of the fact that the natural sciences do, in fact, have unparalleled rhetorical power and prestige in our culture. For example, a recent Gallup poll indicated that scientists were ranked near the top of a list of the most prestigious professions. Although the pollsters did not ask about the relative status of economists, law professors, or literature scholars, they did ask about businesspeople, lawyers, and artists. The comparison was not favorable (Harris Poll 2004; see also Gross 1990, 21; Walton 1997, 15). Indeed, as early as 1969 researchers studying interdisciplinary borrowing found a pattern of upward modeling: "an uncritical selectivity that is overawed in favor of models from disciplines more prestigeful than one's own" (Sherif and Sherif 1969, xii, as quoted in Klein 1990). As John Lyne points out, "perceived competence within a discipline can afford special privileges and immunities, depending on the status of the discipline" (1990, 52).

Before considering in detail the "privileges and immunities" gained by invoking the disciplinary prestige of the natural sciences, I pause to acknowledge the curious fact that writers about science sometimes seek to invoke the prestige of the humanities—not by appropriating the intellectual authority of literary theory (such as it is) but by availing themselves of the cultural prestige of "high art" and literature. Literary scholar Kenneth Knoespel has noted this phenomenon in the enormously popular *Chaos: Making a New Science* by science journalist James Gleick (1987). Gleick not only provides a survey of the scientific research but validates that work by "situating it in the received traditions of the culture," using numerous quotations from great literature to "legitimate the wonders of the new science through canonical expressions of Anglo-American culture" (Knoespel 1991, 105). We see another example of this phenomenon in mathematician Ivar Ekeland's book *Mathematics and the Unexpected*, whose final chapter is devoted to speculations on Homer and Proust (Ekeland 1988, 112–22).[10] Knoespel suggests that scientists themselves sometimes seek validation for their results by situating them in the grand narratives of the culture—not validation as empirically verified, but validation as momentous, profound, and worthy of society's attention and support (1991, 106; see also Hayles 1990, 9). Robert Markley also examines the tendency of some scientists to engage in this curious form of borrowing, and he surveys some recent scientists' invocations of deeper meanings and hidden order. He suggests that contemporary science has inherited from its Newtonian predecessor "its repressed and occasionally half-acknowledged theological imperatives and justifications," and his account could help explain why much of the wildest speculation about the implications of chaos theory for free will, art, and politics was first done by scientists themselves (Markley 1991, 143).

When cross-disciplinary speculation originates in a discipline with relatively more prestige, we can see it as not so much a case of borrowing as of aggressive lending—a bit like taking your new lawnmower over to the neighbor's house and showing them how much better it can cut their lawn. Thus some consider the application of evolutionary thinking to psychology or ethics to be a kind of disciplinary colonization in advance of imperialism. Dupré has raised warnings about just this aspect of some cross-disciplinary work. Although he recognizes that there is nothing wrong with using, for instance, "economic styles of thought as a source of hypotheses about various areas of human behavior," he notes that "typical imperialists do not merely establish embassies in foreign countries and offer advice to indigenous populations. And, similarly, economic imperialists do not merely export a few tentative hypotheses into the fields they invade, but introduce an entire methodology and one that is in many cases almost entirely inappropriate" (Dupré 2001, 128).

Of course, a discipline must have substantial prestige (or at least self-confidence) in order to be able to engage in this kind of cross-disciplinary meddling. I turn now to some of the ways that prestige is harnessed when researchers borrow from chaos theory.

Allies

Sociologist of science Bruno Latour (1987) has provocatively character-ized much scientific discourse as enrolling allies. An embattled researcher can cite respected experts, enlist collaborators, and even appeal to microorganisms and laboratory equipment in order to marshal resources impressive enough to convince the reader. Indeed, the citation of authorities within one's own discipline is standard academic practice in all fields, not just the natural sciences.[11] As noted earlier, we need not accept Latour's stringently rhetorical picture of science in order to recognize that invoking the disciplinary prestige of the natural sciences enrolls entire disciplines as allies. These allies can intimidate people and make it seem as though critics would have to master a new technical area in order to dismiss a contrary suggestion. Sokal and Bricmont consider this behavior an abuse of science when it involves tossing around irrelevant technical terms, with the presumed goal being "to impress and, above all, to intimidate the non-scientist reader" (1998, 4).

The prestige of the natural sciences makes them especially valuable in interdisciplinary settings—perhaps too valuable. In "The Need for Guidelines in Interdisciplinary Meetings," archaeologist Leon Pomerance warned of the problem of peremptory appeals to "scientific data," which may shut down discussion prematurely (1971, 429).[12] Part of the rhetorical power of using technical details resides in their raising the threat of intervention by the "real" experts: "arguments between nonspecialists are thus formulated in such a way as either to escape the opinion of the specialist or to be subject to his decision" and are always affected by the possibility of intervention by the specialist (Perelman and Olbrechts-Tyteca 1969, 104). The rhetorical power of knowledge borrowed from chaos theory often derives less from the technical details of specific findings than from the invocation of powerful allies that seems to hint, "if you sign on with my proposal, look at the additional resources I can bring to our side." Alexander Argyros notes this tendency in the work of two major postmodern theorists, pointing out how Lyotard "enlists such allies as Gödel, Thom, and Mandelbrot" and that "in a typical rhetorical move, Derrida creates allies by simply declaring their compatibility with the deconstructive project" (1991, 234, 34). Demonstrating the power of such allies, Argyros himself enrolls E. O.

Wilson, Douglas Hofstadter, Stephen Hawking, and Marvin Minsky to bolster his conception of evolutionary dynamics.[13]

Seeking to intimidate potential critics by enrolling allies may well seem an illegitimate persuasive technique, whether those allies come from within or without one's home discipline. But just as persuasive citation can play a legitimate role in arguments within one discipline, the cross-disciplinary enrollment of the natural sciences as allies need not be intellectually dishonest. Resonances across disciplines promise both greater intellectual coherence and greater social and cognitive resources. It is not clear that it is necessary, or even possible, to strictly separate the persuasive and the cognitive aspects of these resonances, though it may well be useful to disentangle them in certain contexts. Much of the persuasive power of these resonances can be understood in terms of another rhetorical function of disciplinary prestige: the argument that one's ideas are worthy of elaboration and "pursuit" by other researchers.

Pursuit Worthiness

The disciplinary prestige of the natural sciences can help make a case that one's proposal is worth pursuing. Recall the discussion in chapter 2 of how the context of discovery and the context of justification blur into each other and also meet in a context of pursuit. After achieving audibility, one must convince the audience of listeners or readers that one's proposal deserves further examination. In striving to establish pursuit worthiness, a researcher seeks to make his or her idea into a live option for further exploration, articulation, and testing. And borrowed knowledge has been used to support just this kind of pursuit worthiness in literary, economic, and legal discussions of chaos theory. For example, Peter Francis Mackey contends that chaos offers a "stimulating and resonant opportunity" for new ideas in reading the work of James Joyce (1999, 38). Randall Bausor points out that much of the research on chaos in economics serves "more to illustrate hypothetical possibilities than to compel scientific acceptance" (1994, 119), and Huang and Day explicitly begin their account of chaos in stock prices by stating that their analysis "seems to us to point in a new and fruitful direction for further research" (1993, 169). Not surprisingly, economists prove fond of casting the pursuit worthiness of their ideas in terms of their being smart investments. Thus, Richard Day begins his discussion of a modeling strategy that uses nonlinear dynamics by expressing his hope that "increasing intellectual resources will be invested in its further development," and concludes his treatment of possible policy implications by claiming that "the prospective payoff for investing more intellectual effort in such research would seem to me to be rather high" (1993, 18, 39). Even Royce de R. Barondes, who is intensely critical of the metaphorical use of physics in legal scholarship, acknowledges that although such reasoning cannot establish the validity of its conclusion, it can provide "an interesting line of thought to be pursued" and "may raise productive lines of constitutional analysis" (1995, 174, 176).

How does borrowed knowledge advance one's idea as a candidate for active pursuit? Surely the disciplinary prestige of the natural sciences can help convince an audience that a new approach merits consideration. In some cases, appeals to and connections with other disciplines carry weight by means of the appeal of coherence between theories, a widely recognized epistemic virtue. Because borrowed knowledge offers analogies rather than exact connections, the coherence promised between disparate fields will be of a weaker type than the strict consistency required between, say, biochemical and evolutionary accounts of DNA. But even though an analogical argument may not provide a complete justification for accepting its conclusion, such an argument can raise the conclusion to the level of being a promising hypothesis and thus an active candidate within the context of pursuit. Within this context, where competing active possibilities are ranked according to their promise and feasibility for further investigation, elevating a hypothesis to the status of candidate for active pursuit does indeed count as the conferral of a significant sort of epistemic merit.[14] Here I have been speaking of the use of borrowed knowledge to bolster the pursuit worthiness of a particular hypothesis or an approach to a specific issue or text. In the next chapter I consider at some length the ways chaos theory has been used to support detailed proposals for broad methodological change.

Legitimacy

When researchers fear that their ideas, or indeed their very field of study, will not be taken seriously, they may attempt to invoke intellectual authority in order to establish the legitimacy of their work. Katherine Hayles observes, "No doubt because of the prestige accorded to science within the culture, scientific theories have often been used to validate cultural theses" (1991, 15), and Sokal and Bricmont speculate that researchers "imagine, perhaps, that they can exploit the prestige of the natural sciences in order to give their own discourse a veneer of rigor" (1998, 4). Such attempts may be described disdainfully as trying to make an observation seem profound by "dressing it up in fancy scientific jargon" (Sokal and Bricmont 1998, 11) or as seeking to "cop a certain authority by scientizing" (Livingston 1997, ix). In this section, I focus on some of the strongest criticisms of this use of the disciplinary prestige of science, and in the following chapter I consider some cases in which this tempting resource is employed in appropriate ways.

Sometimes an entire field of study seeks to shore up its intellectual credentials by invoking the disciplinary prestige of the natural sciences. As a telling case in point, Philip Mirowski focuses his indictment of neoclassical economics on his charge that economists made use of a metaphor from physics (utility as a field of force) "in order to appropriate its scientific legitimacy" (1989, 283; see also 357). Indeed, if one wants to make one's field into a science, or at least make it seem more rigorous and less subjective, borrowing knowledge from the natural sciences offers many benefits. Mirowski has also edited a collection of essays, *Natural Images in Economic Thought*, which contains numerous historical examinations of what it has meant in various specific contexts for a particular text to lay claim to "scientific" status. Many of the narratives in this volume explore situations in which economists were engaged in the delicate maneuver of "appropriating the trappings of another science in order to constitute political economy as a separate and self-sufficient science" (Mirowski 1994a, 9). Mirowski further suggests that invoking prestige still occurs in the way that contemporary economists have rushed to embrace chaos: "largely, one fears, because of its popularity amongst the physicists" (1990, 300).

Mirowski, however, sometimes attacks borrowed knowledge a bit too vigorously. For example, he charges that "the conjuration of scientific legitimacy by means of vague innuendo abounds in Samuelson's oeuvre" and cites as an example a 1983 essay by that famous economist (Mirowski 1989, 383). But in that essay Samuelson seems to be seeking to support a certain methodological approach, the relaxing of strict operationalism, by analogy to thermodynamics. There is nothing "vague" about it, for he spends many pages spelling out a detailed isomorphism to physics (Samuelson 1983). Elsewhere Mirowski admits that the physics metaphor has been successful in allowing neoclassical economics to come to be seen as scientific and to become the dominant approach in the United States (1989, 393). Some of the energy for his attacks may derive from Mirowski's strong views about how disciplines achieve legitimacy. For him, inquiries into life, motion, and trade represent interacting points on a triangle, and each rests on the assumption that some special quantity is conserved. Each research program in biology, physics, and economics "derives legitimacy for its radically unjustifiable conservation principles from the homeomorphisms with the structures of explanation at the other vertexes" (116; original in italics). Notice that Mirowski's view makes borrowed knowledge essential, not incidental, because the central principles of each discipline are "purely conventional, and, thus, from a disinterested and detached point of view, simply false" (116). Such an account risks collapsing all disciplines into rhetorical exercises constituted solely by persuasive analogies. But we can criticize particular instances of borrowing-for-legitimacy without accepting a view of research programs as simply floating in the air, held up by coherence with each other's rhetoric.

Drawing on Mirowski's work, attorney Michael Wise has examined a case study of the way the field of law has sought to borrow disciplinary legitimacy. As he puts it, "American antitrust law is an advanced case study of the law's self-denying aspiration to model itself on, and incorporate, the authority of science" (1995, 713). In this case, oddly enough, the science whose prestige is being borrowed is the science of economics, which itself borrowed prestige from physics. Wise contends that "the century of effort to apply science through law, or rather, to do law in the image or shadow of science, teaches that science in law, like everything else in law, is rhetoric. Law mines science for ideas and examples, for labels and platitudes to justify positions in ways that may have little connection with scientists' criteria of justification or understanding" (714). Again, we should be wary of any claim that "everything is rhetoric" in law, or anywhere, because it may tempt us to reduce the legal field to "mere" persuasion. Wise contends that "all of antitrust's borrowings from economics have been analogies and metaphors invoking scientific authority" but does not fully explain whether (or why) these invocations render the analogies illegitimate. Indeed, he makes metaphorical use of chaos theory himself to argue for changes in antitrust policy. Are his analogical arguments nothing more than the conjuring of borrowed prestige? I return to Wise's work in following chapters and explore what, if anything, distinguishes an appropriate metaphor from a manipulative marketing trick.

The Problem of the Halo Effect

The persuasive functions of borrowed knowledge need not be employed inappropriately, but persuasion is not without its dangers. Among the rhetorical problems associated with invoking disciplinary prestige, the distortion introduced by the halo effect deserves special attention. The halo effect is the social-psychological name for the error that occurs when we mistakenly associate traits that do not logically belong together, often by attributing positive traits to an individual, based on the presence of one favorable feature (Walton 1997, 244). The halo effect can apply to individuals or even institutions such as science, in which case we are liable to accept the claims of experts because of the prestige of the institution or discipline to which they belong (Walton 1997, 70). One classic example of the halo effect with respect to personal appearance, delivery style, and apparent professional qualifications is illustrated by the "Doctor Fox Lecture" experiment, in which a lecturer fooled three audiences of mental health and education professionals into taking seriously a nonsensical presentation titled "Mathematical Game Theory as Applied to Physical Education" (Naftulin, Ware, and Donnelly 1973, as quoted in Walton 1997, 152). I contend that part of what contributed to the halo effect in this case was the disciplinary prestige of mathematics. Notice that the infamous Sokal hoax also relied on the prestige of mathematical physics to convince the editors of *Social Text* to publish an article on social and political theory filled with hilariously absurd borrowings from the physical sciences (Sokal 1996).

The halo effect can lead to errors within a discipline, of course, but it can also cause serious problems in cross-disciplinary pursuits. As John Lyne warns, "it is important to be aware of the ways that discourses originating in science may dominate other discourses. When scientific expertise is projected from its home turf onto another, it may be perceived as having more authority than is in fact warranted" (1990, 53). The use of scientific authority to bolster one's argument, or even just one's reputation, brings forth an interesting suggestion from George L. Priest. In criticizing the approach of Clark and Posner, who make use of the science of economics in investigating their particular areas of law, Priest avers that an unhelpful borrowing of scientific theory is being allowed: "Lawyers who practice social science in law schools exploit the successes of true scientists in the underlying disciplines and ride free on their reputation for scientific integrity" (1981, 1293). He goes on to propose that the social organization of legal research needs to be changed to remedy this situation, so that law professors who aspire to make scientific claims should be evaluated by the those in the sciences, not the student editors of law journals. But the phenomenon of borrowed knowledge, with all its attendant problems, does not call for turning over all adjudication of claims about science to the scientists themselves. As discussed further in chapter 5 on metaphor, the criteria for evaluating borrowed knowledge should not be generated or applied only by those from the discipline in which the knowledge originated. Although borrowers ought not "ride free" on the supposed integrity of the sciences, it is not the case that the sciences are the only sources or guardians of integrity in inquiry.

When we reject the argument from authority as a fallacy, sometimes what we mean is that just because someone is an expert in one field does not mean that he or she has any special expertise in another field. Nobel Prize–winning electrical engineers may be utterly ignorant when it comes to genetics. High-energy physicists may know little about cultural theory. As it is with individual expertise, so it may be with entire disciplines. The halo effect can lead us astray when we see the epistemic success of the natural sciences as rendering their concepts and methods as especially helpful for solving problems in other fields. Sokal and Bricmont have some useful advice for avoiding this problem: *"Don't ape the natural sciences,"* they urge. "The social sciences have their own problems and their own methods; they are not obliged to follow each 'paradigm shift' (be it real or imaginary) in physics or biology" (1998, 187, emphasis in original). This sound advice reflects an admirable disciplinary pluralism: the natural sciences have valuable knowledge and methods to offer, but we need not assume that their techniques can answer every question we have.

The disciplinary prestige of the natural sciences is largely well earned, based on their tremendous theoretical and practical successes. Invoking their prestige can be a legitimate and appropriate rhetorical maneuver. But to at least some extent, the natural sciences are accorded unearned prestige and intellectual authority because of the halo effect and other cultural factors that lead us to defer to scientists and consider their form of knowledge far superior to any other. To just that extent, invoking the prestige of the natural sciences runs the very real risk of reinforcing their "surplus prestige" and solidifying cultural attitudes that denigrate artistic, humanistic, or spiritual inquiry. As Lyne points out, the dangers of the halo effect open the theoretical question of how discourse domains should relate to each other and to the public domain (1990, 53). I do not attempt to solve this problem here but return to it in chapters 7 and 8, where I explore the use of scientific knowledge to address practical questions of value. For now, let us return to one of the most important rhetorical functions of borrowed knowledge: its use in motivating large-scale changes in methodology.

4 Motivating Methodological Change

An investigation of borrowed knowledge from a rhetorical perspective has highlighted some of the less savory uses of chaos theory, such as flavoring one's work with irrelevant technical terms and intimidating opponents with the prestige of the sciences. But borrowing has other roles to play, especially in arguments for broad changes in a discipline's methodology. Such arguments can never be fully supported by existing evidence, so researchers find that they need to construct "rational conversion" narratives (Gross 1990, 98; see also van Eemeren et al. 1996, 346). Attempting to convince one's disciplinary colleagues to adopt a new approach requires more than simply stating facts; it requires doing the work of persuading an audience that the new approach is worthy of pursuit and deserving of their time, effort, and resources. Inquiry does not involve simply operating the machinery of research to churn out new knowledge; it requires a researcher to undertake a particular practice—where "practice" must be understood as including not only beliefs and assumptions but standards of behavior, ideals of explanation, and skill at particular techniques. Convincing an academic to adopt a substantially new methodology may involve just the kind of persuasion that Thomas Kuhn (1970) described as playing a crucial role in establishing a new "paradigm."

And borrowed knowledge can serve as a powerful tool for accomplishing this rhetorical goal.

This form of persuasion involves reasoning, not simply good salesmanship. Even Philip Mirowski, as harsh a critic of borrowing from physics as one could ever find, admits that the attempts by classical and Marxian economists to emulate physics were "not simply a case of envy or misplaced obeisance (although one can't rule that out *tout court*)—after all, was there a better paradigm of effective causal explanation close to hand?" (1989, 186). In offering an argument for methodological change, a researcher may give a reason based on how others in different fields have succeeded in doing things the new way. Such an argument relies on an analogy between the two fields of study, and it may involve borrowing images, concepts, techniques, and results. The disciplinary prestige of the field that is the source of the analogy will strengthen the appeal, because asking someone to make the leap to a promising but undeveloped approach is indeed asking a lot. If one can appeal to the success of a similar approach in a different and highly successful field of inquiry, the leap of faith required may well seem less daunting.[1] Chaos theory has brought a wide variety of new phenomena within the scope of scientific understanding, enlarging the range of behaviors amenable to mathematical analysis. These developments have also challenged and enlarged the conception of what counts as a genuinely scientific understanding, providing a powerful resource for those wishing to redirect their own disciplines and perhaps even claim a share of the disciplinary prestige accorded to the natural sciences. In this chapter I examine some detailed cases of the use of chaos theory to motivate change, especially in economics but also in legal and literary studies. I argue that in several instances this use is perfectly appropriate and begin by sketching the terrain in the study of chaotic economic behavior.

Chaos and Economics

Chaotic behavior was available for study, theoretically and experimentally, for decades before it became popular in the 1980s. I have discussed elsewhere some of the reasons for this delay—reasons both technical (lack of computing power) and cultural (a social interest in prediction and control) (see Kellert 1993, chap. 5). It appears that some of the same reasons kept economists from exploring the possibility of chaotic behavior in economic systems, at least until chaos theory became widely popularized in other fields. For example, Nobel Prize–winning economist

Ragnar Frisch discussed chaotic behavior in economics as early as 1933, but his efforts received little attention (Lonçã 2000).[2]

One of the most important reasons for the delay in investigating chaos was a prejudice in favor of studying linear systems. Although nonlinearity was treated by some earlier economists, for most of them their approach was generally linear and seeking equilibrium. Although "every significant economic process and institution involves nonlinearities, through much of the history of economic theory in general and business cycles in particular has been an attempt to work around nonlinearity for reasons of analytic tractability" (Mosekilde et al. 1993, 60; see also Day 1993, 22). For example, in the study of business cycles most economists were fixated on linear equations, but such equations simply cannot generate behavior rich enough to model persistent economic oscillations—they can only settle down to stasis or explode. Some early work by Goodwin noted that nonlinear functions can yield persistent oscillations, but this is where things were left, with no one exploring ranges of parameters that could generate chaos (Baumol and Benhabib 1989, 79). Thus, there was no way to see how cycles could arise in simple systems except by means of exogenous shocks. Those who talked about nonlinear systems generating intrinsic cycles were mostly ignored, until recently (Goodwin 1993b, 304; see also Kelsey 1988, 2). I call this a linear prejudice because the preference for the random shock model, which assumes that fundamental economic relationships are linear, is not based on economic evidence. Instead, the preference results from the fact that this model was the first one proposed and that it uses statistical techniques commonly taught to economics graduate students (Kelsey 1988, 21).

What made linear systems so appealing? Part of the answer has to do with the fact that they are especially amenable to solutions that yield stable, predictable equilibria. Traditional economics has been described as emulating the regularity of physical systems: "The Newtonian paradigm underlying classical and nonclassical economics interpreted the economy according to the pattern developed in classical physics and mechanics, in analogy to the planetary system, to a machine and to clockwork; a closed autonomous system ruled by endogenous factors of highly selective nature, self-regulating and moving to a determinate, predictable point of equilibrium" (Weisskopf 1983, as quoted in Prigogine 1993, 4). As Randall Bausor points out, one cannot explore chaotic dynamics without paying attention to instabilities. "By inclination and training, however, economists abhor instability. . . . To most economists competitive processes that rule the economy are inherently dynamically stable" (1994, 122). And that is why economists model randomness with exogenous shocks. "Their most cherished attitudes toward markets and their most central presumptions about how the economy should be governed are all profoundly challenged by analyses conditioned on systematic instability" (Bausor 1994, 122). One of the goals of this chapter is to examine how these challenges can be bolstered by borrowed knowledge.

As mentioned in chapter 1, investigations of economic chaos began to take off in the 1980s. One early example of the application of chaos theory within microeconomics deals with a model for the economic adjustment process in a system with a single good (this discussion follows the treatment in Jensen and Urban 1984). Here, the descriptively named "market clearing equation" gives us a deterministic dynamical system for the price of the good. If the supply and demand curves are linear, only simple types of price behavior can occur. However, with the introduction of a strong nonlinearity, the price behavior can undergo a well-understood transition to chaos. Such a nonlinearity can arise when the demand curve "bends," that is, when demand for the good actually increases as the price goes past a certain point. What this (highly simplified) mathematical model shows is that such nonlinearities can result in extremely complicated chaotic price behavior.

By 1993, Richard Day could claim that chaotic behavior had been shown to be present and robust in many economic models, including price mechanisms and business-cycle theories: "in short, chaos is generic in virtually any model of a dynamic economic process that retains inherent nonlinearities" (1993, 23). These simple economic models "can generate complex behavior qualitatively like real world experience," meaning that they show prices and macroeconomic indices fluctuating irregularly (24). It is crucial to note that these nonlinear models show how economic variables can display complex behavior "on the basis of internal forces alone, with no resort to unexplained outside influences" (28).

In macroeconomics, it had long been thought that random fluctuations must be caused by external stochastic shocks.[3] But work in the 1980s by economists such as Benhabib and Day (1982), Grandmont (1985), and Boldrin and Montrucchio (1986) produced models of irregular business cycles that still respected economists' cherished assumptions about "rational" behavior and perfect markets. This work revived interest in the possibility of endogenous fluctuations (Dechert and Hommes 2000, 652). For example, uncertainty about the plans of rivals, customers, or regulators can give rise to random fluctuations in a market, even when all economic "fundamentals" such as inflation and gross domestic production behave regularly (Cass and Shell 1989, 4). Later work expanded on the early results to explore which conditions can be relaxed and still allow for chaotic dynamics in the models: what degree of competition, what timescales, and so forth. These chaotic model systems range from models of advertising expenditure versus net profit (Baumol and Benhabib 1989, 99) and capital accumulation (Mosekilde et al. 1993) to "overlapping generations" models in which consumer choices change based on increasing wealth, past choices, or the choices of other consumers (Kelsey 1988, 11) and growth models that assume that economic actors have limited information or "bounded rationality" (Day 1993, 32; see also Woodford 1989).

Models of economic chaos have faced serious criticism, however. For example, Brock and Malliaris point out that theoretical models of overlapping generations don't work too well when the most common length of business cycles is 3.5 years, quite a bit shorter than human generations (1989, 323). Proving the existence of chaos in a model may seem to do little more than establish it as merely a theoretical possibility. Indeed, Michael Woodford has said that the possibility of chaos is a fantasy created by "theorists bearing free parameters undisciplined by empirical studies" (1987, as quoted in Rosser 1991, 119; see also Day and Chen 1993, 319). And Philip Mirowski heaps scorn on what he calls the "little mathematical exercises" of "single-variable matchbook models" that are supposed to account for apparent economic randomness (1990, 301).

In response to this challenge, economists have attempted to show that their models do not make outlandish assumptions or unreasonable choices of parameters. And they have searched for empirical evidence of nonlinearity and chaotic behavior in real-world economic systems. Turning to the first of these responses, we find an admission that early models were highly simplified and not very realistic (Kelsey 1988, 2). But recent work has endeavored to show the presence of chaos in models with more plausible assumptions and empirically relevant parameter values (Nishimura and Yano 1995; Mukherji 1999; Lagos and Wright 2003; Cellarier 2006). And recent work on higher-dimensional models as well as approaches that incorporate work on learning and complex systems approaches has revived interest in chaotic models (Dechert and Hommes 2000, 653). An especially interesting example of this recent work involves rational agents choosing between expensive but accurate tools for predicting prices and cheaper but less effective ones. The switching between these two strategies creates the kind of stretching and folding of trajectories that is responsible for chaotic dynamics (Brock and Hommes 1997; this model is further generalized in Goeree and Hommes 2000). This work moves beyond the one-dimensional models of earlier work and represents an interesting hybrid of the strange-attractor approach of chaos theory and the adaptive-agents approach associated with complexity theory, coupling the dynamics of a market to the process of choosing reliable predictors of that market.[4]

Questions remain, however, as to the real-world implications of these theoretical results. In their survey of chaos in economics, Brock and Malliaris point out that natural scientists will not be impressed by a mathematical model that proves the bare possibility of a type of behavior. They will ask whether the parameter values needed are empirically warranted and whether the behavior is actually found (1989, 299). It is in part the urge to emulate natural scientists that has driven economists to search for signs of chaos in actual data. Researchers have attempted to find strange attractors in such records as commodities futures prices or a nation's gross domestic product, and a "fierce debate" sprang up over the interpretation of the results (Chen 1993, 217). Brock and Malliaris characterize this debate as turning on whether a model based on plausible economic behavior can account for the time series data better than, say, a rational expectations model with stochastic shocks. Although no one believes that macroeconomic time series are generated by a strictly deterministic dynamics, the question is whether one can find a low-dimensional attractor underneath some of the ever-present economic noise (Brock and Malliaris 1989, 299–305).

One point seems clear, however: economic data show signs of nonlinearity.[5] Nonlinear behavior has been reported in rates of return of stocks (Hinich and Patterson 1993, 201) as well as an aggregate measure of the money supply (Barnett and Hinich 1993, 261). More recent articles have announced a "broad consensus of support for the proposition that the (macroeconomic) data generating processes are characterized by a pattern of nonlinear dependence" and "clear evidence of nonlinear dependence" in financial data (Barnett and Serletis 2000, 715; see also Dechert and Hommes 2000, 652). This kind of nonlinearity is necessary but not sufficient for the presence of chaotic behavior. Barnett and Chen enthusiastically announced their results in a 1988 article titled "The Aggregation-Theoretic Monetary Aggregates Are Chaotic and Have Strange Attractors: An Econometric Application of Mathematical Chaos" (see also Barnett and Choi 1989, 154), but elsewhere the results have been less resounding. In some cases, when the time series data from macroeconomic systems are subjected to data-analysis techniques used to find intrinsic low-dimensional chaos, no such strange attractor is found (Scheinkman 1990). This negative result leaves open several possibilities: the conventional explanation may be correct, or the data may merely be too limited to allow the strange attractor to be detected, or the attractor may be of very high dimension, or some explanation using the concepts of complexity theory may suffice (Bak et al. 1993). A recent survey of 19 published articles looking at data from stocks, currencies, gold, and silver found 12 positive reports of nonlinearity, but in the search for chaotic behavior there were four sightings and four results in which no chaos was found (Abhyankar, Copeland, and Wong 1997, 2–3).[6] The evidence for deterministic chaos in economic and financial data appears to be weak (Dechert and Hommes 2000, 652; Shintani and Linton 2004; Serletis and Shintani 2006), but a more recent study found evidence of chaos in data from energy futures (Matilla-Garcia 2007). One article found no chaos when testing for sensitive dependence on initial conditions, but concluded that this "cannot be considered as a definitive answer" and summarized the debate by saying that "the currently available evidence on chaos in the macroeconomic time series is rather ambiguous" (Shintani and Linton 2003, 332, 346).[7]

Part of the reason for this unresolved state of affairs is that the search for empirical evidence of chaos faces many difficulties. For one thing, it is difficult to gather data of the quantity and quality that natural scientists enjoy because the time series is typically shorter and noisier (Barnett and Choi, 1989, 152; Chen 1993, 223; Rostow 1993, 17). These relatively short time series may make low-dimensional structure appear where it isn't, and the tests for chaos do not give unambiguous results when looking at noisy data (Barnett and Hinich 1993, 255). On the other hand, some diagnostics may reject deterministic chaos even when it is there. And sometimes the government steps in, shifting the rules of the economic system and making long-term data hard to interpret (Brock and Malliaris 1989, 311, 328).

In fact, a number of economists have concluded that it is downright impossible—a "virtually unsolvable problem"—to determine whether a real-world set of economic data displays chaotic behavior or not (Barnett and Hinich 1993, 255; see also Brock and Malliaris 1989, 322). In addition to the problems caused by errors of measurement and aggregation, and the amplification of noise, there is no way to distinguish chaos resulting from internal dynamics and chaos due to shocks driven by an external chaotic system such as the weather (Dechert and Hommes 2000, 653; Barnett and Serletis 2000, 721; Barnett 2006; see also Mirowski 1990, 304). Indeed, scientists can not always empirically determine whether a stochastic model is appropriate even in physical systems

(Leiber 1998, 368). So we are left with an unresolved situation. Dechert and Hommes suggest that "the truth behind the real forces of business cycles is probably somewhere in between the two alternatives, nonlinear endogenous forces buffeted by exogenous shocks" (2000, 660). This pluralistic approach may seem unsatisfying to those who want a definite yes or no. But however the dispute about evidence for economic chaos turns out, it is clear that borrowing from the natural sciences has spurred an active program of research at least for now.

Methodological Change in Economics

Although economists still disagree about the applicability of chaotic dynamics, methodological change may nonetheless be motivated by appeal to the availability of actual computational techniques. In the context of active disagreements, where one cannot yet convincingly demonstrate the success of one's approach, it is crucial to establish at least its plausibility. The prestige of physics can help to establish the pursuit worthiness not only of a particular idea (as discussed in the previous chapter) but also of a broader method or even an entirely new approach. For example, William Barnett and Seungmook Choi compared two different approaches in econometrics, conventional statistical methods and the techniques of searching for chaotic attractors. They began their survey by recognizing that because the traditional methods are not intertranslatable with the techniques of chaos theory, and that they cannot substitute for each other, a question arises as to whether a pluralistic attitude toward these different approaches is warranted: "It is natural to ask whether the investment of time necessary to incorporate these new tools [such as fractal dimensions and strange attractors] into econometrics would be profitable" (Barnett and Choi 1989, 141). They concluded that the chaotic-attractor approach was worth pursuing because it can produce "theoretically deep" explanations and "powerful results" (142, 151). This conclusion rests partly on the strength of their empirical analysis and partly on the disciplinary prestige accorded to physics on account of its "depth" and "power."

It may seem that the disciplinary prestige of physics is irrelevant here: after all, economists can prove theorems about their models and provide precise measurements for the likelihood that their statistical analyses are valid. Might it not be that economists borrow the techniques of chaos theory, but that the source of the techniques—the context within which they were discovered—matters not one bit when it comes to justifying their results? Consider a comparatively straightforward example from one of the earliest appearances of chaos in the economics literature, the 1986 work of Michele Boldrin and Luigi Montrucchio. Many economic models assume that agents will behave in regular, predictable ways, and this behavior is often held to be due to their "maximizing" behavior—always seeking rationally to pursue the greatest payoff of any decision. Boldrin and Montrucchio construct a standard model from capital theory, called an optimal growth model, wherein they demonstrate that optimizing agents nonetheless exhibit chaotic behaviors (1986, 26–27). This result seems valuable, for it challenges a basic assumption about the regularity of economic agents, and the authors state that they consider their results "the opening to a potentially fruitful field of research" (37).

Why only *potentially* fruitful? Is there more at work here than polite academic understatement? A clue may be found in Randall Bausor's investigation of why chaos theory has been well accepted in fluid dynamics but not in economics. In fluid mechanics, a system can be characterized by the Navier-Stokes equations, which are regarded as authoritative. But in economics there is nothing even close in terms of authoritative systems of equations. So many different alternative foundational equations are available that any particular model seems ad hoc (Bausor 1994, 121). Yet the Lorenz equations, which serve as one of the clearest exemplars of chaos theory, are in fact derived from the Navier-Stokes equations by means of such a drastic series of approximations and simplifications that they must be seen as something of a metaphor for fluid flow rather than an exact model (Bausor 1994, 117). Physics has authoritative models (however inexact or metaphorical their application), whereas economists must continually strive to portray their models as plausible. In that effort, surely the disciplinary prestige of physics can be of aid.

Faced with a throng of competing models and a long history of seeking to emulate physics in order to establish scientific credentials, some economists have found in chaos theory a powerful ally indeed. To convince one's colleagues to make a radical departure from established approaches, it doesn't hurt to be able to point to examples from fluid dynamics and laser optics. And if the techniques being proffered are brand-spanking new, so much the better, given the rhetorical power of "newness" described in the previous chapter. In the following section, I briefly touch on some directions of methodological change for which economists have enlisted chaos theory: (*a*) a move away from modeling randomness in terms of external shocks; (*b*) a willingness to explore nonlinearities in economic equations; (*c*) a reconception of the role of economic models; and (*d*) a redirection toward economic dynamics as opposed to equilibrium. But first, I briefly discuss some of the more radical challenges to established economic methodology.

The dominant form of economics in this country is the approach called "neoclassical." Enshrined in the standard textbook by Nobel Prize winner Paul Samuelson, this approach builds on the calculus-based techniques of the nineteenth-century marginalists and emphasizes the way markets reach equilibrium—where supply and demand balance at a stable price or savings and investment gravitate toward a self-perpetuating level of national income (Samuelson 1980, 57, 211). One of the hallmarks of traditional economics is the efficient markets hypothesis. As Samuelson explains, "what looks much like a random walk is just how stock or commodity prices *should* look in an *efficient market*. Competitors discount in advance all that can be confidently foreseen and anticipated. So any good news clearly ahead is *already* in the IBM price. And so too is any anticipatable bad news ahead. What, then, makes stocks move? It is the arrival of *new* news—rain on the Kansas wheat fields, loss of an antitrust court case by IBM. These chance events make speculative prices vibrate randomly—wobble up and down" (1980, 70, emphasis in original). As we will see, the notions that economic systems do not settle down to stable equilibrium, and that erratic economic data are due to the intrinsic dynamics of a chaotic system, challenge the efficient markets hypothesis and with it the neoclassical consensus.

Opposition to neoclassical economics has come from Keynesian, Marxian, and institutionalist approaches, and these alternative approaches have found some affinities with chaos theory. Thus, Rosser states that neo-Keynesian and neo-Marxist models can show chaos but that neoclassical economists "have tended to assume away these results and to cling to any model that will produce convergence to some well behaved steady state equilibrium" (1991, 321). From other heterodox positions, Irene van Staveren claims that chaos theory can provide institutionalist economics with illustrative metaphors that aid in intuitive, visual understanding (1999, 164; see also Radzicki 1990), and Buchanan and Vanberg cite nonlinear dynamics in support of the notion of radical freedom they find in "subjectivist economics" (1991, 178). But I will be focusing on more mainstream economic traditions.

One noteworthy challenge to the neoclassical consensus addresses precisely the issue of borrowed knowledge. This challenge comes from historian of economics Philip Mirowski, whose book *More Heat Than Light* (1989) ignited a lively controversy about the ways economists have used the concepts and techniques of nineteenth-century physics. It is well established among historians that the model of the natural sciences played a crucial role in the development of mathematical economics (Porter 1994, 128). What is not so clear, however, is how well these economists understood what they were borrowing. For Mirowski, their errors render neoclassical economics fundamentally incoherent.[8]

Mirowski holds that chaos theory supports his criticisms of neoclassical economics, and he draws on the work of Benoit Mandelbrot, who has studied the fractal structure of markets. Neoclassical economics is based on differentiable, continuous functions, but Mandelbrot found that prices evolve with radical discontinuity, nowhere differentiable (Mirowski 1990, 296; see also Mandelbrot and Hudson 2004). According to Mirowski, chaos and strange attractors are important for economics as part of an overall acceptance of intrinsic randomness and discontinuity. But he insists that they cannot be simply added to neoclassical economics: "If followed to their bitter conclusions, chaos models would render orthodox theory meaningless" (1990, 304–5), ultimately requiring us to see an intimate blurring of order and chaos incompatible with neoclassical determinism. However, I do not enter the debate about the legitimacy of neoclassical economics, because my project primarily focuses on the ways that borrowed knowledge is used within the fields of economics, law, and literature as they are currently configured in the academy.

Moving Away from External Shocks

Returning, then, to conventional economics, recall that traditional stochastic macroeconomic modeling considers a system subject to utterly random external (or "exogenous") shocks, whereas the chaotic-attractor approach constructs a fully deterministic set of equations that nonetheless "can move in a seemingly arbitrary and unpredictable way" (Goodwin 1993a, 48). Richard Goodwin, one of the pioneers of the use of nonlinear dynamics in economics, says that the possibility of intrinsic (or "endogenous") irregularity "seems to me to require a reformulation of some econometric procedures," including both forecasting and planning governmental interventions (1993a, 49). A number of other economists have similarly argued that the failure of the stochastic approach to economic forecasting, and the success of deterministic approaches in such fields as meteorology, suggests a reexamination of the theoretical foundation of econometrics (see, e.g., Chen 1993, 245). In this way, an appeal to the legitimacy of chaotic models in physics helps to argue for a new approach to understanding contemporary economic dynamics.[9]

Moving Away from Linear Prejudice

Linear models have long dominated the theoretical modeling of business cycles, even though Richard Goodwin introduced a nonlinear cyclical model as early as the 1950s. His model was fully deterministic and did not rely on outside forces to drive the cycles, but it had the drawback of producing cycles that repeated in exact, regular patterns. Researchers interested in nonlinear dynamics have argued that one of the important lessons of chaos for economists is that linear models can be misleading and that linear theory "is not an appropriate foundation for the study of economic cycles" (Bullard and Butler 1993, 865; see also Mosekilde et al. 1993, 59). "For business cycles modeling, one of the criticisms of the early post-war research in nonlinear dynamics was that the actual business cycles are not exactly repetitive, as the deterministic model would imply.... However, the discovery that complicated, random-looking sequences might prevail in deterministic models has dampened that criticism substantially. Therefore, it is really the possibility of chaos that revived interest in nonlinear dynamic models" (Bullard and Butler 1993, 853).

The prospect of understanding economic fluctuations in terms of intrinsic dynamics rather than external pushes and pulls has helped to attract attention to aspects of economic systems that had long been passed over. Many of these neglected features depend on nonlinearities in the underlying equations, and, as we have seen, nonlinearity is a prerequisite for the possibility of chaos. Here we find another way in which chaos theory has helped motivate a redirection of economic methodologies: it "stimulates the researcher to look for magnification or overreactive effects in data in the search for evidence of the endogenous instability that is characteristic of chaotic dynamics" (Brock and Malliaris 1989, 339). Effects such as the magnification of small fluctuations, which occur when people overreact to new information, are essentially disproportionate and thus characteristically nonlinear. Chaos theory can help to legitimize the systematic treatment of such previously disregarded behavior (see Cunningham 1994, 571).

Brock and Malliaris provide a plausible account in which these nonlinear behaviors can help explain the market phenomenon of speculative fads (1989, 336). Although abstract mathematical models in economics may seem far removed from rhetorical concerns, each of them contains a story. The dynamical behavior of the model introduced by Brock and Malliaris, for instance, is explicated in terms of a narrative in which fads arise, grow, and dissolve. The prestige of nonlinear dynamical models in physics can help legitimate the consideration of factors that play a role in these economic stories—factors that can generate chaotic dynamics in economic models. Thus, factors such as irrationality, time lags, and differences in generations or styles of investment are opened up as worthy of pursuit.

Moving Away from Exact Predictions

In their 1989 survey of the significance of chaos theory for economics, William Baumol and Jess Benhabib note that chaos makes economic forecasting very challenging because the two main forecasting methods are thrown into question: first, extrapolation becomes a doubtful strategy because chaotic systems can exhibit the behavior known as intermittency, in which they seem to settle down into a regular oscillation but then have another pattern suddenly emerge; and second, constructing a model with estimated parameters becomes difficult if a small change in a parameter can have drastic consequences for the model. In response to these challenges, Baumol and Benhabib recast the purpose of chaotic models in economics: "the work on chaotic dynamics suggests that disenchantment with earlier dynamic models is perhaps attributable to failure to recognize their most promising role—that of revealing sources of uncertainty, and enriching the list of recognized *possible* developments" (1989, 80).

This exhortation recalls Boldrin and Montrucchio, mentioned earlier, who proved that economic models do not necessarily work the way they had been assumed to work. Although their sets of equations are far from claiming the authoritativeness of chaotic systems in fluid dynamics, they should not be dismissed as mere toys. Their purpose is not to generate predictions of actual economic systems but to expand our conception of what kinds of things economic systems can do and of what kinds of economic factors merit study. In motivating such a methodological reorientation, it can be helpful to harness the prestige of sciences such as fluid dynamics and even meteorology, which are considered fully "hard" sciences while still being notoriously unable to deliver reliable predictions. The disciplinary prestige of these natural sciences that have let go of any insistence on exact predictability can help ease the transition to a new methodological approach.[10]

Moving from Equilibrium to Dynamics

Because traditional neoclassical economics focuses on situations that have already reached equilibrium, it has been said to happen in a realm without time (Murphy 1994, 558). In a scene of economic equilibrium, it does not matter how we got there, and all transactions are perfectly reversible. Such an economics has no genuine dynamics to it, no ir-reversible changes, no *history*. Yet nonlinear dynamics opens up the possibility of systems that evolve in irreversible ways, introducing such elements of genuine historical change as hysteresis (see Kellert 1993, 93–96). So strong is the ahistorical character of traditional economic equi-librium that Mirowski has predicted that it would "block the wholesale embrace of chaos theory" were economists ever to understand it fully (1989, 390). The emphasis on equilibrium holds so much power in eco-nomics that Joseph A. Schumpeter once characterized the existence of a unique equilibrium point as nearly a requirement for any exact science (Arthur 1990, 93). Yet chaos theory offers a way for some economists to challenge the conventional notion of economic equilibrium as a timeless, static state.

If one is committed to linear models, it is natural to think that the only sensible long-term behavior of a system—the only "equilibrium" worth considering—must be a steady state devoid of change or evolu-tion. James Bullard and Alison Butler have criticized the way economists often impose strict stability conditions on their models because of a com-mitment to the linearity of those models. In linear models, dynamical paths that do not settle down to a steady state usually seem not to make sense economically because they would either explode or rely on spe-cial parameter choices (1993, 851). But, as discussed earlier, nonlinear models open up new possibilities for reasonable models that display endogenous fluctuations: their failure to settle down to a steady state is not because they are subjected to external shocks. Bullard and Butler identify such a possibility as "important for economic theory" and say it "seems quite promising," making a pitch for its pursuit worthiness (1993, 853; see also Wise 1995, 762–63). In their account, it no longer makes sense simply to assume that the only economic equilibrium is a steady-state situation, because nonlinear dynamics invites us to concep-tualize long-run economic equilibria as any kind of attractor—steady states, cycles, or chaos. Now, cycles and chaos do not qualify as "equi-libria" in the mathematical sense (an unchanging state of all variables), but they match the traditional conditions of an economic equilibrium such as clearing the market. This is an important distinction, and it marks an important challenge to traditional economic methodology. It certainly changes the connotation of the word "equilibrium" in eco-nomics.

Although Bullard and Butler admit that there is not yet a "completely convincing demonstration" that chaos can exist in nonlinear economic models with plausible parameter values, the power and the prestige of nonlinear dynamics in the natural sciences strengthens the case for broadening the conception of economic equilibrium. David Kelsey echoes this point in stating that "perhaps one of the greatest services dynamic systems theory has done is to make economists aware that other types of solutions are possible" besides steady states and cycles (1988, 26). In questions about the meaning of economic equilibrium, as in questions about the role of external shocks, nonlinearity, or prediction in economic models, knowledge borrowed from chaos theory can serve as a valuable resource for motivating methodological change.

Methodological Change in Law

Although the previous discussion of nonlinearity and equilibrium in economics at times turns on technical points, in disciplines such as law, the use of chaos theory often involves developing metaphorical arguments for very general methodological precepts. Attorney Michael Wise observes that when we seek to understand the challenges posed by chaos theory, "technical details are less important than the philosophical and cultural implications of changing expectations about what science can do" (1995, 752). In arguments for change in the practice of legal scholarship, the reasoning often takes the following form: the target field is like the source field in certain relevant respects; therefore, methods of inquiry that succeed (or fail) in the source field can be hypothesized to succeed (or fail) in the target field. In this way, methodological injunctions, warnings, or comforts issue from the metaphorical translation involved. In this section I explore two examples: first, the contention that legal reasoning need not derive predictions from exact principles, and second, the attempted redirection of legal theory toward a "dynamical" approach similar to that discussed earlier in economics. In making such arguments, legal scholars often take issue with traditional conceptions of scientific methodology that have informed current thinking about how legal studies ought to be conducted, and appeal to both the newness and the prestige of recent work in the natural sciences.

Moving Away from Exact Predictions

Many of those who study the law, as well as those who make it, have long sought to give to their enterprise the rigor and objectivity that supposedly attaches to scientific research (Veilleux 1987, 1967; Williams 1987, 430). A long-standing conception of science, rooted in the tremendous success of Newtonian physics, pictures scientific knowledge as enabling powerful and accurate predictions based on exact and clearly specified physical laws. And Christopher Columbus Langdell, the Harvard Law dean who revolutionized American legal education, firmly believed that "through scientific methods lawyers could derive correct legal judgments from a few fundamental principles and concepts" (Grey 1983, 5). As a result, Jason Scott Johnston characterizes one of the main concerns of recent jurisprudence as "the ability to formulate legal rules that yield certain or at least predictable outcomes" (1991, 341; see also 344n4), and Edward Imwinkelried notes that a faith in scientific certainty is "deeply embedded in American Evidence law" (1995, 56).

Although both ends of the political spectrum have emulated scientific predictability, political conservatives are among the strongest voices insisting on predictable judicial outcomes, including Judge Richard Posner, whose law and economics movement envisions the law as a reliable device for facilitating the smooth functioning of business. Posner states in the founding issue of the *Journal of Legal Studies* that the goal of legal studies is "to make precise, objective, and systematic observations of how the legal system operates in fact and to discover and explain the recurrent patterns in the observations—the 'laws' of the system" (1972, 437). Similarly, Robert Bork's constitutional jurisprudence "demands a powerful predictive ability as the test of legitimacy for constitutional theory" (Reynolds 1991, 113, quoting Bork's *The Tempting of America*). Law professor Glenn Harlan Reynolds (familiar to many from his "instapundit" blog) contends that proponents of the left-wing critical legal studies movement also hold that only legal principles that yield predictable results are worthwhile. In the face of such widespread importance placed on predictability, Reynolds argues that chaos theory may be "an especially useful source of analogy because it suggests an important parallel between constitutional scholarship and modern physics—the degree to which both are engaged in a quest for 'grand theories' that will explain events without contradictions or messy uncertainties" (1991, 110).

Reynolds draws an analogy between the evolution of legal principles and the behavior of an exemplary chaotic system such as the dripping faucet: in both situations, an element may expand until it reaches a critical limit and then break off to begin the cycle again. On the basis of this analogy, he states that "unlike scientists, who have learned better, most legal scholars still expect to be able to predict when and how that will happen—and consider any theory that will not do so inadequate" (Reynolds 1991, 112). For Reynolds, chaos theory calls into question the methodological injunction that we must look for predictability in particular cases and suggests that we look instead for the underlying order corresponding to strange attractors. And even if we fail to find such order, at least chaos theory can help to clear away outdated notions of how science works (Reynolds 1991, 116; see also Wise 1995, 757).[11]

Reynolds suggests that in surveying the patterns of decisions made by the Supreme Court, for example, chaos theory can help us to steer between the twin perils of rigidity and nihilism: he notes that despite being unpredictable, "the actions of the Supreme Court are not random. Just as there is structure within chaos, so there is pattern of sorts within the actions of the Court—pattern that itself reflects recursion and sensitivity to initial conditions, and that exists on both large and small scales" (1991, 114). Although allowing that the Court isn't actually "chaotic" in the same strict sense as a physical system, Reynolds argues that chaos theory nonetheless calls into question the idea that the prediction of particular cases should be a goal of legal studies (116). In this, he draws a powerful lesson from nonlinear dynamics—that exact quantitative prediction is not always possible, so science can learn much from attending to larger-scale patterns and underlying order (see Kellert 1993, chap. 4). If even physics need not obey Posner's methodological strictures, why should legal studies? To quote Hayek, "the methods which scientists or men fascinated by the natural sciences have so often tried to force upon the social sciences were not always necessarily those which the scientists in fact followed in their own field, but rather those which they believed that they employed. This is not necessarily the same thing. The scientist reflecting and theorizing about this procedure is not always a reliable guide" (1952, 14).

A number of legal scholars have made use of chaos theory to help motivate a similar methodological shift. As Wise suggests, "the law may be relieved to abandon the delusive goal of predicting particular outcomes, for its focus has always been more on rules, principles, and institutions anyway" (1995, 764). Indeed, J. B. Ruhl suggests that full-blooded legal positivists who want legal theory to take full advantage of modern science should in fact be thankful for chaos theory because now they can give up the search for a completely predictive legal theory (1996, 928). And Imwinkelried notes that the Supreme Court in the *Daubert* ruling (*Daubert v. Merrell Dow Pharmaceuticals, Inc.* 113 S. Ct. 2786, 7794 [1993]) accepted a notion of scientific testimony that acknowledges the loss of certainty in modern science. He cites chaos theory as one of the reasons to reject a faith in scientific certainty, which has major implications for all of evidence law (Imwinkelried 1995, 78). In each of these cases, the prestige of physics helps attract attention to the call for change, to motivate readers to adopt a new methodology in legal study, or to cushion the shock that may be experienced in relinquishing traditional practices.[12]

Moving from Equilibrium to Dynamics

We can also find a number of legal scholars who suggest that chaos theory can serve as a model for reorienting our thinking about law, toward a more dynamic approach, similar to the methodological urgings described previously with regard to economics. Michael O. Wise draws a lesson for antitrust law from the challenge to the notion of a static equilibrium, illustrating a link between economic theory and legal policy. For Wise, the law has followed economics in its fixating on static equilibrium states rather than dynamical processes of economic evolution.[13] He urges that "the focus of economic explanation and legal policy should shift from these comforting delusions about imaginary end-states to the structure and evolution of institutions, to the patterns and expectations that guide and move economic behavior in a world that is not, and never will be, in equilibrium" (1995, 740). Wise urges a methodological shift similar to some of those described earlier in the discussion of nonlinearity: by relaxing the insistence on steady-state equilibrium, economists could be free to "explore more realistic conceptions about the knowledge and motivation of economic actors and about their interactions with each other" (760; see also Day 1993, 21).

Jason Scott Johnston speaks of the "dynamics of legal change" in constructing a detailed dynamical model of unpredictable oscillations between bright-line rules and balancing standards in torts decision-making (1991, 346). For Denis Brion, chaos theory "suggests that law might be better described in terms of process rather than structure" so that we need not seek an optimal end-state of legal doctrine (1995, 193). And Andrew Hayes (1992, 757) offers the dynamical approach of chaos theory as a refutation of those who would claim that science is "too rigid, too simple, to provide a model for the dynamism and plasticity of society that a system of jurisprudence must reflect." A more extended use of chaos theory as a model for legal dynamics comes from Robert Scott, who suggests that the pattern of legal decisions handed down by the courts is like a strange attractor in that the tension between the need for present justice in a particular case and the need for future justice in similar cases generates dynamic swings between the two poles of the "justice paradox." The history of such shifts, he claims, "is like the infinitely recurring patterns produced by a swinging pendulum. The patterns are not identical, but are different in scale or size. So, too, the recurring patterns of contemporary legal thought are not simply the re-cycling of new ideas, not merely the pouring of old wine into new casks. They are like fractals. Each pattern is similar to the past, but different in scale" (Scott 1993, 350). In each of these cases, as with those who argued against the fixation on equilibrium in economics, chaos theory provides a useful tool for motivating methodological change.

Methodological Change in Literary Studies

Borrowed knowledge can also support a much broader methodological appeal—an invocation of the objectivity of the physical sciences. Legal scholars may see in chaos theory a way to supplement the perceived validity of their enterprise, and humanities scholars may see the use of concepts from physics as mitigating the perceived fuzziness of liter-ary interpretation. In fields where researchers feel a lack of objectivity, or sense that their discipline receives less respect because of its seem-ing subjectivity of method, borrowed knowledge can provide the sheen of objectivity that attaches to scientific knowledge.[14] Although it may strike some as odd to speak of "methodology" in the study of litera-ture, we can find examples of chaos theory being used in attempts to change the way we read and interpret fiction. Consider the way Gordon Slethaug invokes chaos theory in order to urge readers to pay attention to small changes within texts and their ramifications through "itera-tions" that are both like and unlike what came before. He remarks on anticipations of this advice: "Literature, of course, has been telling readers that before science gave the process its own vocabulary, but in literature . . . differences or discrepancies are often minimized and all but edited out by the emphasis on generalizable behavior, but stochastics and chaos theory show how even the tiniest difference or error can, af-ter several iterations, create huge discrepancies" (Slethaug 2000, 125). Beyond providing a technical vocabulary for this suggestion, nonlinear dynamics also lends its considerable disciplinary prestige to Slethaug's project of arguing for a different kind of close reading than that of the formalists or New Critics.

Harriet Hawkins provides another example of the use of borrowed knowledge to support methodological change in literary study, making a sustained argument that chaos theory can help move us away from forced and one-dimensional readings of literature. She targets "com-paratively linear critical, aesthetic, moralistic, and ideological ideals of order" that yield impoverished readings (Hawkins 1995, 5; see also 18).

Hawkins claims that because tragic heroes and heroines suffer so terribly, "linear-minded moralists" have tried to find in them proportionately terrible sins or flaws. But chaos theory shows how trivial events can bring about disproportionately momentous effects (16). Hawkins recognizes that everyday life is full of such "butterfly effects" and that they have been a staple of literature for ages. What, then, is the use of borrowing the image of sensitive dependence on initial conditions? It is surely not needed to confirm our everyday knowledge of disproportionate effects. One wishes that Hawkins had provided some actual examples of these ideologically rigid critics—the moralists and "literary determinists" who insist that the unfolding of a tragic plot must follow a straightforward, inevitable, and rigorously proportional path. However, we should see her as making a methodological argument about how to go about reading texts: pay attention to the uncertainties and the details that complicate the story beyond a simple tale of justice being paid out or power structures being enforced. If there really are critics who insist on imposing strict "big effect–big cause" proportionality or rigid categories on every text, then chaos theory may in fact be a useful tool for motivating an audience of literary scholars to reject such interpretive methods. And there is nothing wrong with using the disciplinary prestige of science to advance this aim: if one wants to show that good knowledge can be made without doing something, and chaos theory shows that even physicists do not need to do that thing to accomplish their work, then it makes a lot of sense to use metaphors from chaos theory.[15]

Hawkins makes a similar methodological argument for the viability of comparing works from very different cultural contexts, based on the fact that the mathematics of chaos can be used to investigate very different physical systems. Here, she argues against "an *exclusive* concentration on literature's cultural, ideological, and historical roots and branches" and "all past and present taboos against treating together, as structurally comparable (rather than historically or categorically noncomparable), ancient and modern and popular and canonical works of varying genres that show the same pattern" (1995, 47). Although again she does not offer any examples of critics insisting on such exclusive concentration and taboos against such comparisons, the analogy may still be a good one. If some knowledge in the physical sciences can be fruitfully transported across contexts, perhaps some aspects of literature are not entirely fixed by their contexts either. Hawkins shows how the rhetorical strategy of invoking disciplinary prestige can be both effective and appropriate.

Conclusion

In this chapter and the previous one I have shown how a rhetorical approach can help us understand some of the functions served by borrowed knowledge. Indeed, the rhetoric of science is one of the few fields that has directly addressed the phenomenon of borrowing. For example, Dilip Gaonkar points out that the rhetoric of science has studied "how and why and with what effect 'findings' in one discipline are analogically translated into 'premises' of another discipline," generally moving from "harder" to "softer" sciences (1997, 40). Lyne has an excellent description of the borrowing of knowledge in his discussion of sociobiology. He identifies the way the rhetoric of sociobiology calls attention to "the potential for borrowing rhetorical resources from one domain and using them in another. Such a strategy makes biological considerations discursively available to discussion of social, political, or moral issues, or vice versa—the flow can be in either direction." In this cross-fertilization, "heuristic resources from one discursive domain enter and animate another" (Lyne 1990, 38). This portrayal is exactly the inventive use of borrowing.

So, having looking at the persuasive uses of borrowed knowledge, I now turn to their uses in invention—specifically, the cognitive role of metaphors from science in generating and justifying new ideas outside the sciences. However, because we have called into question the distinction between the contexts of discovery and justification, can we really separate the persuasive from the cognitive? Perhaps not, and yet perhaps a strict separation is not necessary for us to identify different questions. If someone claims that the pattern of Supreme Court decisions can be thought of as a strange attractor or that a certain novel has a fractal structure, we may well ask What does this metaphor do for us? or Why should we accept that? The process of answering these questions need not be isolated from considerations such as the following: Who is called on in support of this claim? Why is this claim even worth listening to in the first place? and How well would this way of looking at things fit with other parts of my knowledge? In gaining audibility, recruiting allies, and invoking disciplinary prestige, an inquirer seeks to persuade, yes; this endeavor is but one part of the process of making knowledge.

In a naturalist (and, I would add, pluralist) spirit, Philip Kitcher has argued that empirical research in the sociology of science can make an important contribution to epistemological inquiry. By examining concrete examples of the impact of social relationships on knowledge production, we can gain valuable insights that will help us discern which social arrangements help generate the best knowledge (Kitcher 1998, 46–47). Of course, this normative project will be impossible unless we can resist the imperialist urge to declare that all knowledge is merely a reflection of power relationships. This chapter and the previous one make possible a parallel argument about the rhetorical investigation of cross-disciplinary work. We should not simply discount the role of persuasion in knowledge construction; neither should we collapse all inquiry into "mere" rhetoric. Instead, we can and should look for evidence about which persuasive techniques promote the development of knowledge, as well as about the risks and dangers that they can pose. In this spirit, I turn now to an investigation of another element of the process of knowledge production that has often been discounted: the role of figurative language in general and metaphor in particular.

5 Metaphorical Chaos

The inventional use of borrowed knowledge is described by historian of science I. B. Cohen as not a direct transference or exact copying but as a "creative transformation"—"an intellectual leap forward that often happens when a concept, a method, a principle, or even a theory is transferred from one domain to another" (1994, 66). In this chapter I investigate the creative transformations made possible by the metaphorical use of scientific knowledge. First, I briefly examine how metaphor works, in order to outline the ways it can function in the transfer of knowledge across disciplines. Then, I argue that such metaphorical borrowing can serve a useful purpose, and I conclude with general remarks on the curious relationship between chaos and metaphor. Here, as in the next chapter, my main target is the belief that metaphorical borrowing cannot or should not be subjected to detailed investigation. Among those who hold this belief are some self-styled defenders of science who claim that any metaphorical use of science is always an inappropriate misuse. For example, Paul Gross and Norman Levitt simply declare that chaos theory has nothing to do with broad cultural questions (1994, 105; see also Weinberg 1996, 12; Pigliucci 2000, 66). Other writers, such as Alexander Rosenberg (1994) and Maurice Lagueux (1999), consider metaphors to be merely inspirational devices confined to the "context of discovery" and thus unimportant as targets for criticism. And some, such as Richard Rorty (1991), view metaphors as mysterious triggers for radical conceptual change, devoid of cognitive content and unsuitable for theoretical investigation. In contrast to all these views, I consider metaphorical uses neither so freewheeling as to be immune from criticism nor so wrongheaded as to be unworthy of criticism. Both sweeping condemnations of metaphors and blanket licensing of them fail to pay adequate attention to the way metaphor actually works in the transfer of knowledge across fields.

What Is Metaphor?

Three Theories

The great variety of theories of metaphor can be roughly grouped into three basic positions: the dismissive view, the comparison view, and the interaction view. Holders of the dismissive view think of metaphor as merely a special trick with words—perhaps useful as a persuasive device or heuristic source of inspiration but worth no serious attention in the pursuit of knowledge. Some trace this view back to British empiricists such as John Locke (cited in Kittay 1987, 5; see also Haack 1998, 69–71), George Berkeley (cited in Gentner and Jeziorski 1993, 448), or Samuel Parker (cited in Lakoff and Johnson 1980, 191), writers whose fierce denunciations of figurative language occasionally employed their own colorful metaphors to hilarious effect. The empiricist tradition insisted on linguistic precision, which required explicit definitions before using a term. The dismissal of metaphor was taken up by positivists, who viewed metaphor as having only emotive and heuristic use (Kittay 1987, 7). Andrew Ortony characterizes contemporary holders of this view as insisting that metaphors are violations of linguistic rules and serve rhetorical ends but contribute nothing to the making of knowledge: "rather unimportant, deviant, and parasitic on 'normal usage'" (1993, 2).

A second view holds that a metaphor may in fact accomplish something in the pursuit of knowledge, but that it is entirely replaceable by an equivalent statement, usually the presentation of a comparison. So if I say "Man is a Wolf to Man," that simply means that humans are similar to wolves in their viciousness. This view is associated with the philosopher of language Donald Davidson, who has claimed that metaphors have no special meaning, serving simply to direct our attention to some similarity between two things, just as the corresponding simile does (1984, 256).

The position I adopt in what follows is called the interaction view, associated with the philosopher Max Black. According to this view, metaphors can have a distinctive cognitive function and content, which cannot be replaced by a literal statement of comparison. Metaphor takes the structure of associations and relationships that surround the "source" field (wolves, in the example) and brings some of that structure to bear on part of the "target" realm (here, humans).[1] These two fields interact so that "the wolf-metaphor suppresses some details, emphasizes others— in short, *organizes* our view of man" (Black 1962, 41). Kittay summarizes the main features of the interactionist account by saying that a metaphor is an utterance with two components in tension, where the irreducible cognitive meaning of the metaphor arises from the interplay between these two components understood as systems (1987, 22–23).

How Interaction Works

An interaction metaphor can be used to bring some order and structure to the target field, or to subtly change our thinking about it. Structuring metaphors can bring additional richness of order and relationships to a field that lacks them.[2] So the relationships that we know apply to temperature (for example, hotter and colder, with an absolute zero at one extreme and lukewarm in the middle) can be applied to performance in basketball (for example, they're really hot; they're really cold; they're frozen solid; etc.) Consider the example Donald Schön offers of industrial designers coming to see a paintbrush as a pump: a designer of synthetic bristles said the metaphor "led him and the other researchers to notice new features of the brush and of the painting process" (1993, 141). The metaphor projected notions associated with pumping onto painting, transforming the designers' perception of it. New elements came into the foreground: for example, the space between the bristles came to be seen as a channel. The designers could then pay attention to new elements such as the angle of bend, and this attention led to new explanations and inventions.

In an interaction metaphor, as the philosopher Nelson Goodman describes it, "a whole set of alternative labels, a whole apparatus of organization, takes over new territory. What occurs is a transfer of a schema, a migration of concepts, an alienation of categories" (1968, 73). When we describe a battle as a chess match, the field of chess is "projected upon" the field of warfare: "the chess vocabulary filters and transforms: it not only selects, it brings forward aspects of the battle that might not be seen at all through another medium" (Black 1962, 41–42). A change in emphasis accompanies this new perspective, and the relationships and roles appropriate to board games are now read onto the battlefield. Thus, we find ourselves thinking about warfare in a new light, and the interaction means that our thinking about chess will change as well. Such metaphors can be expressed succinctly in a short phrase or spelled out in detailed analogical schemes (Fahnestock 1999, 24, 41).[3]

The interaction view describes the role played by metaphor both within science and in borrowings from science. Richard Boyd provides a valuable example of how the metaphor of the mind as a computer is used within cognitive science to give a direction to a young field of research. The metaphor structures our thinking about the mind so that we seek out its different parts: the one responsible for memory, the one responsible for processing, and so on for each smaller mental task. We then think about problems with the mind as problems with information processing. As Boyd formulates it, these theory-constitutive metaphors also play an important role "in the development and articulation of theories in relatively mature sciences" (1993, 482, 494). They function to introduce new theoretical terms and initiate investigation into the properties of these newly identified features of the target field. As an example of the metaphorical transfer of chaos theory outside the natural sciences, recall Robert Scott's suggestion that the pattern of legal decisions handed down by the courts can be thought of as a strange attractor. For Scott, the tension between the need for present justice in a particular case and the need for future justice in similar cases generates dynamic swings between the two poles of the justice paradox. This metaphor brings the features of a strange attractor, and its associations and differences (it is different from mere noise, or a simple clock; it has pattern and structure but also unpredictability), to the field of legal decisions. I return to this metaphor later to evaluate it in terms of the effectiveness of the interactions it induces.

Defense of Metaphor: How Does Metaphor Work Well?

General Defense

I begin a brief defense of metaphor against various claims of its ir-relevance or eliminability by sketching four lines of theoretical defense, before proceeding to my main argument, which is concerned with examples of metaphorical transfer. But allow me to say a word or two about some of the philosophical issues I will be passing over. I do not offer a detailed argument against philosophers such as Davidson and Rorty, who state that metaphor has no special meaning beyond some pragmatic use, because this issue has little bearing on my project. Whereas Rorty wants to assimilate metaphors to surprising nonlinguistic phenomena that cause us to rethink our webs of beliefs and desires (1991, 167), Davidson (1984) admits that metaphors can bring analogies and similarities to our attention. Birdcalls and supernovae cannot do this sort of thing. So it is worth asking how, in general, metaphors bring such features to our attention, and how our attention is directed—not in a mechanically predictable way, but in a way that is nonetheless intelligible. Rorty also draws a divide between the interaction view and Davidson on the question of whether metaphors can serve as justifications for belief (1991, 169). My concern, however, is not whether a metaphor can itself be a reason for a belief, but whether we can make sense of the cognitive process by which a metaphor helps us to form different beliefs. So much work in literature, law, and even economics concerns the construction of ways of reading, seeing, and working that the traditional philosophical language of inductive reasoning and justification of isolated beliefs is not helpful. Davidson admits that theories of metaphor can tell us about the effects that metaphor has on us, and this agreement is all that is necessary for my project.

Another set of issues I am not going to worry about is whether or not metaphors are true, or "really refer," or successfully give us correct knowledge of the way the world really is. In this I follow Max Black, who declines to get involved in disputes over metaphorical truth (1993, 39–41). After all, Boyd makes an argumentative leap from the epistemic success of theories that employ metaphors to their epistemic access to a real truth (1993, 519–20), but Lakoff and Johnson make metaphor the basis of an extended argument against objectivist epistemology (1980, chaps. 25–27). And Kuhn can agree with Boyd concerning almost all that he has to say about the use of metaphor in science, and the epistemic success of science, while still not adopting his ontological realism (Kuhn 1993, 541). These questions about semantics and ontology have a metaphysical flavor to them: Is there really a Truth independent of all human activity that metaphors can latch onto? Are there really things Out There that metaphors successfully name? And how do they successfully represent or refer to them? I do not intend to use the word "metaphysical" as a term of abuse, but such questions do not deal with the central issue of my project.

And what is that central issue? Just this: How can the results of science be used metaphorically in successful nonscientific knowledge-making endeavors? Even if metaphors never actually succeed in referring to things or justifying truths but just guide us in a roundabout way to whatever kind of knowledge it is that we can get, then this issue still requires attention. My main concern in this chapter and the next centers on the usefulness or inadvisability of making metaphorical jumps, and the question of this usefulness or inadvisability cannot be solved at the level of abstract ontological theorizing.

Returning now to the defense of metaphor, one starting place is the claim that metaphor is simply a practical necessity. Goodman, for instance, insists that far from being merely ornamental, metaphor provides an economy of expression that we could not do without: "Metaphor permeates all discourse, ordinary and special, and we should have a hard time finding a purely literal paragraph anywhere. In that last prosaic enough sentence, I count five sure or possible—even if tired—metaphors" (1968, 79). Finding a literal paraphrase (of the metaphorical "sadness" of a painting, for instance) would be unwieldy to say the least (93).

One of the crucial practical purposes of metaphor is its rhetorical function, as described in chapters 3 and 4. Sometimes terms from chaos theory are used metaphorically for motivating illustration (to put it nicely) or as glamorous jargon (not so nice). For example, the poet Samuel Taylor Coleridge explained that he attended Sir Humphrey Davy's lectures on chemistry "to renew my stock of metaphors" (as quoted in Barth 1995, 329), and Justice Benjamin Cardozo employed the concepts of relativity and quantum physics "as a literary device to symbolize and make vivid the relations among people, social institutions, and legal rules and concepts" (Veilleux 1987, 1978). Such illustrative uses have a rhetorical force that some would characterize as merely decorative, motivational, or persuasive, and lacking any cognitive content, but this depiction would draw too sharp a line between the practices of persuasion and the production of knowledge. As I showed in chapter 3, the distinction between the persuasive and the cognitive should not be conceived as a rigid split.

Black indicates a second line of defense when he characterizes metaphor as an indispensable conceptual (or, we might say, "inventional") resource. He defends the usefulness of metaphor in terms of the resources it makes available for stretching our concepts: we need to make use of this flexibility, "the available literal resources of the language being insufficient to express our sense of the rich correspondences, interrelations, and analogies of domains conventionally separated; and because metaphorical thought and utterance sometimes embody insight expressible in no other fashion" (1993, 33; see also Langer 1957, 141). Indeed, Black goes so far as to defend what he calls the strong creativity thesis:

"some metaphors are what might be called 'cognitive instruments,' indispensable for perceiving connections that, once perceived, are *then* truly present" (1993, 37). According to this view, "a metaphorical statement can sometimes generate new knowledge and insight by *changing* the relationships between the things designated" (35). Although this view may seem to raise red flags with regard to the issue of realism and truth, Black clarifies the sense in which metaphors bring into being new relationships: "Did the slow-motion view appearance of a galloping horse exist before the invention of cinematography? Here the 'view' is necessarily mediated by a man-made instrument.... And yet what is seen in a slow-motion film becomes a part of the world once it is seen" (37). The strong creativity thesis is in fact a stronger defense than what is needed simply to justify attention to the cognitive function of metaphor, so I do not make further use of it.

It is, however, important to defend the interaction view against the comparison or "replacement" view of metaphor. Proponents of that view, such as John Searle (1993, 111), claim that the so-called metaphorical interaction adds nothing to a mere comparison. Yet Mary Hesse pointed out, in her early and influential work on models and analogies in science, that the function of a living metaphor is always open-ended. Any attempt to reduce it to a statement of comparison between already-explicit literal descriptions robs it of its distinctive purpose (Hesse 1966, 162–63). As Kittay asks, with reference to a poem by Wallace Stevens, "What objectively given similarity is there between a garden and a slum prior to the formation of the metaphor itself?" (1987, 17). To return to Donald Schön's example, the mapping of pump onto paintbrush did not happen right away—elements and relations had to be regrouped and renamed in order to see *A* as *B*. The designers certainly did not start by noticing similarities. There was an inchoate sense of similarity, and then they articulated the elements in their restructured perceptions to formulate an analogy. In the final stage of the process, one may be able to take a step up into abstraction and see both source and target as instances of a general model (Schön 1993, 142–43). The comparison view may indeed be correct about the function of metaphor in some contexts, but the interaction view more adequately accounts for many of the uses of borrowed knowledge that I consider. As with disciplinary approaches, we can be pluralists with regard to our theories of metaphor.[4]

A third line of theoretical defense would question the line between the literal and the metaphorical. Mary Hesse points out that the interaction view of metaphor challenges any fixed distinction between literal language as rule following and figurative language as rule breaking: "Intelligible metaphor also implies the existence of rules of metaphoric uses" (1966, 165). Goodman insists that the metaphorical applicability of a predicate is no more mysterious than its literal applicability, because there is always a "jump" being made (1968, 78; see also Fahnestock 1999, 22). Kuhn points out that theoretical terms and even observational terms are open-ended in their application and unable to be cashed out in clear-cut necessary and sufficient conditions, just as metaphor is (1993, 534). There is an element of metaphor even in the use of the quantitative analytical methods of chaos theory, for instance, because a lot of these techniques live in the land of abstract phase space. Kuhn points out that the metaphor of atom as solar system was replaced by the more detailed Bohr model, but applying the model still requires seeing similarities and teasing out where the model fits and where it doesn't: "I would hazard the guess that the same interactive, similarity-creating process which Black has isolated in the functioning of metaphor is vital also to the function of models in science" (Kuhn 1993, 538; on this point, see also Hesse 1966; Giere 1988). So even within the natural sciences, the line between the literal and the metaphorical becomes blurred. When "carrying over" scientific knowledge to a new domain, similarities must be defended as strong enough to justify using familiar techniques in new places. For example, in the move to economics, we note that there is (as yet) no laboratory that allows one systematically to alter parameters and observe changes in the behavior of entire stock markets or national economies, so Randall Bausor contends that all models of economic chaos remain "wholly metaphorical" (1994, 123; for an even stronger view, see McCloskey 1985, 75).

The fourth defense is even bolder and claims that the metaphorical is in some sense prior to and more fundamental than the literal, that "language is always already metaphoric" (Hayles 1990, 34; see also Lakoff and Johnson 1980, 121). The linguistic resources we develop through metaphor may at some later point collapse into literality, but down the road we will need metaphor anew—so metaphor comes both before and after the literal. We will see another version of this defense in chapter 9, where it appears in the often disparaged guise of deconstruction, but such a characterization of all language as metaphoric is stronger than what is required for serious attention to metaphorical uses of borrowed knowledge.

What if all these defenses fell through? What if every metaphorical statement was in principle replaceable by a statement of comparison or an encouragement to notice a particular similarity? After I describe some of the metaphorical uses of chaos theory, the reader may look and see whether these metaphorical borrowings were dependent for their usefulness on the very open-endedness and suggestiveness that make literal paraphrase difficult if not impossible. Even Davidson agrees with Black, Goodman, and others about what metaphor accomplishes, but he simply denies that it accomplishes this by means of a distinctive meaning or other cognitive content. He explicitly disavows the view that metaphor is "confusing, merely emotive, unsuited to serious, scientific, or philosophic discourse. . . . Metaphor is a legitimate device not only in literature but in science, philosophy, and the law; it is effective in praise and abuse, prayer and promotion, description and prescription" (Davidson 1984, 246). My project is precisely to seek out the modes and the limits of that effectiveness.

Specific Defense

The core of my more specific defense of metaphor follows Eva Feder Kittay's description of its inventional function—its role in generating hypotheses from current conceptualizations and transforming those conceptualizations. Juxtaposing perspectives yields a reconceptualization, bringing new properties to salience, and reorganizing concepts (Kittay 1987, 4). Indeed, even as Sokal and Bricmont warn that transference between very different fields of thought can be used to hide weaknesses, they admit that "the observation of a valid analogy between two existing theories can often be very useful for the subsequent development of both" (1998, 11). Let us then proceed with investigating the specific uses of metaphor.

Most inventional uses of borrowed knowledge as a metaphor can be divided into two broad and overlapping categories: metaphors that structure and those that restructure. As Kittay explains, a metaphor creates a mapping of the structure of relations from part of the source field onto part of the target field. If the target field lacks a rich structure, a structuring metaphor *induces* structure upon part of it, and previously unorganized elements of the target field will take on the organizational structure of the source field. If the target field is already highly structured, a restructuring metaphor *reorders* part of it, temporarily or permanently, by bringing the elements of the target field into a new relationship that matches the structure of the source field (Kittay 1987, 170; see also Lakoff and Johnson 1980, 61). Now, "mapping" is a technical term in mathematics, borrowed in turn from geography. Mathematical mappings between sets of elements can take many forms, one of the strictest being an isomorphism in which the two realms very nearly mirror each other. But Kittay points out that in an effective metaphor, the mapping need not be isomorphic, but it must be relation preserving or "homomorphic" (Kittay 1987, 169; see also Gentner and Jeziorski 1993).[5] So in mapping temperature terms onto basketball skill, it doesn't matter that one scale has more terms than the other; what matters is that the relation of graded antonymy is preserved.

Structuring metaphors provide valuable services for an inchoate field of inquiry: they can introduce theoretical terminology for important features and generate hypotheses for possible pursuit.[6] In these ways, the conceptual scaffolding provided by the source field can enable the systematic investigation of new phenomena. Recall the discussion in chapter 2 of the different stages of inquiry; once we accept that the generation, pursuit, and testing stages are not walled off from each other, it is easier to grant a genuine role for metaphor in the production of knowledge (Haack 1998, 81). Economist Giorgio Szegö states that use of techniques from physics by early economists "had the positive effect of inducing a certain quantification in economic thought and allowing the emergence of more generally valid theories, the derivation of more realistic models, and most importantly, the formulation of more relevant, correct, and precise questions" (Szegö 1982, as quoted in Mirowski 1989, 386).[7] Although we might be tempted to consider the unlocking of new approaches to be a wholesome and positive use of borrowed knowledge, Ira Livingston sounds a more cautious tone, saying, "any theory (e.g. 'chaos theory') that seems to legitimize categories of phenomena otherwise excluded from consideration also opens up, by this validation, new realms to disciplinary penetrations" with a "dynamics of simultaneous liberation and colonization" (1997, 3). Inducing structure in order to enable systematic inquiry might not always be a welcome development.

One of the ways in which a structuring metaphor opens up new areas for research is by providing theoretical terms for a phenomenon. Recall Richard Boyd's discussion of irreplaceable metaphors that are used to express theoretical claims for further investigation. He offers as examples the claims "that thought is a kind of 'information processing,' and that the brain is a sort of 'computer,'" "the view that learning is an adaptive response of a 'self-organizing machine,'" and "the view that consciousness is a 'feedback' phenomenon" (Boyd 1993, 488). Boyd notes that "even among cognitive psychologists who despair of actual machine simulation of human cognition, computer metaphors have an indispensable role in the formulation and articulation of theoretical positions. These metaphors have provided much of the basic theoretical vocabulary of contemporary psychology," and "their cognitive content cannot be made explicit" (487). In economics, Neil Niman has argued that even if metaphors from a natural science such as evolutionary biology cannot do much to help make economic predictions, they play an important role in helping formulate more useful and accurate descriptions of economic events: "Using concepts borrowed from evolutionary biology may be of value by imposing a structure on the thought process that forces distinctions to be drawn that would otherwise go unrecognized because they are not required by a theory of price" (1994, 379).

Beyond the introduction of a theoretical vocabulary, structuring metaphors can guide the development and refinement of scientific hypotheses. Boyd speaks of the "programmatic research-orienting" feature of metaphors in terms of their ability to invite the reader "to explore the similarities and analogies between features of the primary and secondary subjects, including features not yet discovered, or not yet fully understood" (1993, 489; see also Mirowski 1989, 277; Ortony 1993, 255). Recall the discussion of the paintbrush as pump. This metaphor helps generate pursuit-worthy hypotheses: if we think of the paintbrush as a pump, then it might be worth seeing whether the amount of pressure influences the rate of flow, for example. This inventional function of generating a new hypothesis is complemented by the persuasive function of convincing an audience that it is worthy of pursuit. As we have seen in chapter 2, although the former function seems to take place within the "context of discovery" and the latter within the "context of pursuit," the line between these contexts cannot be drawn sharply.

Structuring metaphors generate conjectured inferences that are candidates for testing. Although the "fit" or structural soundness of the mapping can be determined by examining the relationships for correspondence, checking whether the inferences are factually correct is a different process (Gentner and Jeziorski 1993, 452; see also Boyd 1993, 494; Fahnestock 1999, xi). Such investigation might not pan out. Thus, some see an important role for metaphor in science but confine it to the "context of discovery" rather than the "context of justification." Perhaps this move draws inspiration from the economist Alfred Marshall, who proclaimed, "analogies may help one into the saddle, but are encumbrances on a long journey" (as quoted in Limoges and Ménard 1994, 336).

Turning next to the restructuring function of metaphors, Kittay provides the example of a poem by John Donne that not only points out similarities between courtship and fishing but induces conceptual relations between lover and beloved that mimic the relations found in the source field (1987, 275). The restructuring metaphor leads us to reconceive the target field, and if the restructuring is fully accomplished the metaphor may lose its lively force (286). Andrew Hayes appeals to the restructuring power of metaphor when he introduces his article on chaos and law with the claim that "by providing a way to understand an order in nature that had previously been seen only as confusion, chaos provides the basis for a new theory of jurisprudence that both explains the nature of the legal process in an entirely new way and re-connects jurisprudence to a natural philosophy that encompasses both science and the humanities" (1992, 752). In what follows, I focus on three particular functions that restructuring metaphors can accomplish: they can defamiliarize fields whose structure has become rigid; they can serve as an antidote for previous metaphorical importations; and they can introduce "third terms" that disrupt existing dichotomous structures.[8]

As a first step toward restructuring, a metaphorical borrowing can shake up the existing structure of a field, inducing researchers not to take its organizing assumptions for granted by defamiliarizing stagnant assumptions. Legal scholar Laurence Tribe describes how our vocabulary lags behind our evolved understandings, so our questions and answers are still expressed in a way that does not reflect our new perceptions. Therefore, "interdisciplinary comparison brings greater awareness of preconception, and it is the unearthing of such tacit knowledge that often creates the possibility of choice and intellectual progress" (Tribe 1989, 3). Reynolds points out that even if metaphorical explorations do not help us learn new and interesting things about the law, they can at least help clear away old ideas about scientific methodology and free us from models that have been abandoned by physicists themselves (1991, 116–17). As Andrew Hayes puts it,

> chaos theory is relevant to jurisprudence because chaos rejects the underlying assumptions of what nature is "like" which, without even being noticed, are used as the conceptual models for how we think about law. . . . The metaphor of chaos differs from other theories only in that it is visibly so, because it is so different from the prevalent ways of seeing and doing law. That is, chaos is avowedly metaphorical because we have forgotten to "see" our current theories as metaphorical structures, only noticing their existence when a new theory questions using them as the basis for the "normal" way to describe things. (1992, 756–57, 765)

In the study of literature, William Paulson claims to practice interdisciplinarity by importing terms from information theory or self-organization in order to "disturb, enrich, and perhaps displace the study of literature by injecting into it some information sufficiently foreign" to serve as "noise" and shake up existing patterns of explanation (1991, 49).

Defamiliarization can open up the possibility of new models or, as Michael Wise puts it, new values: "the system of scientific ideas the law has borrowed does not model reality realistically and does not represent well values that legal institutions have found it important to vindicate. The system's greatest strength is its familiarity. That very familiarity will make it more difficult to change" (1995, 771). For Wise, defamiliarization is the first step toward a second, more substantive restructuring function: the replacement of outmoded models in the target field. Following the work of Philip Mirowski, he traces the importation of conceptions of atomistic individuals and equilibrium-seeking systems into economics from nineteenth-century physics, and he goes on to lament the importation of these independent rational maximizers into legal thinking. According to Wise,

> legal analysis rests now on this 19th-century image of science; what would happen if that image were brought up to date? . . . If antitrust remains warmed-over 19th century scientific metaphor, it will not survive into the 21rst. Antitrust law must either abjure scientific pretenses or at least adopt a new scientific metaphor. This article examines approaches that a new metaphor might take, still premised on linking law to science, but making that link to a different scientific or mathematical tradition or understanding. (1995, 715–17)

The chain of importations carries a risk: the target field can be stuck with a sclerotized conceptual framework insulated by the leftover authority of a now-debunked view: "Economics' deliberate emulation of physics closely paralleled law's borrowing of economics, each claiming the prestige of someone else's ideas. Justification by authoritative analogy breaks down when the source of the borrowed ideas changes its mind" (Wise 1995, 730; see also Hayes 1992, 765). In this way, a new use of physics both "unfreezes" the older usage and proposes a new direction: an "antidote" for a previous importation.

Mirowski, upon whom Wise depends heavily, admits that the status of the original model in the source field doesn't matter—an economic view based on outmoded physics may still yield correct economics (Mirowski 1989, 398). Oddly enough, D. A. Walker makes this same point in his critical review of Mirowski, in which he states that no matter where its metaphors came from, the important thing about neoclassical economics is the practicality of the tools it gives us (Walker 1991, 623). Klein identifies a problem with borrowing ideas that are "out of favor in their original context (including an overreliance on 'old chestnuts')" (1990, 88, drawing on Pomerance 1971). But if that previous importation gave rise to a useful structure in the target field, there is no reason to abandon it just because the source field has changed its mind. Nevertheless, the antidotal function of a metaphor can undo the legitimation of a previous importation to the extent that such legitimation was the result of the rhetorical power of the prestige of the source field.

Metaphors can induce many kinds of new structures, often by breaking up rigid dichotomies. Restructuring takes the form of dichotomy busting when the target field had been structured in terms of a binary opposition, a strict dichotomy such as randomness or planning, utter mess or clocklike regularity. Chaos introduces a third term in physics between total randomness and rigid order, and a metaphorical importation from chaos theory can serve to restructure another field that also had a strict dichotomy between structure and randomness, offering a third option. The restructuring produces three terms instead of two, bringing a new conceptual organization to an existing field. Rhetorical theorist Jeanne Fahnestock describes the tactic of creating composite names as a useful means for undoing oppositions: "managed competition" in health care and "constrained contingency" in discussions of the origin of life serve to create new options that escape from rigid dichotomies and have the luster of paradox:

> Whether perceived as paradoxical or not, a bridging alternative can be looked at either as the ground that absorbs or unites opposites in a larger category or as a composite alternative. But a third term that mediates between antitheses can participate in still another kind of undoing. The fundamental argumentative tactic here amounts to presenting an audience with three things at a time instead of two things at a time. Two things can always be set in opposition, but three things, provided they are perceived as on the same "plane" or a part of the same grouping, cannot be. This third thing can simply be set "next to" the other two, and the net result can be a kind of flanking move on an antithesis. (Fahnestock 1999, 89)

By characterizing chaotic behavior as "deterministic randomness," physics offers a valuable, seeming oxymoron that outflanks the dichotomy

between order and randomness by embracing both of the opposing
terms.

Metaphor as Chaos

The interaction theory of metaphor helps us to understand some of the
ways in which scientific knowledge such as that offered by chaos theory
can be used metaphorically in other fields. An existing field of knowledge
can induce a structure upon an otherwise unordered field of inquiry—
providing valuable terminology and provoking hypotheses worthy of
investigation. Metaphorical borrowing can also bring about the reorganization of an existing structure, calling into question previous importations or introducing alternative conceptual possibilities. In the
next chapter I examine how knowledge borrowed from chaos theory
can fulfill all these metaphorical functions, but I also show that not
all metaphorical uses are equally effective or worthwhile: we can and
should critique them. So my examination of metaphor yields a threefold
defense of their potentially useful cognitive function in the transfer of
knowledge across disciplinary lines: a theoretical defense of the inelimin-
ability of metaphor, a specific outline of its actual function, and a scheme
for sorting such uses into better and worse. Why develop a defense so
manifold that it may seem to be overkill? Because the cognitive legitimacy of cross-disciplinary borrowing faces such a serious challenge:
metaphor seems to be threatening, even dangerous, to its critics and even
to some of its defenders.

   In my earlier book, I described some of the resistance to the investigation of chaotic phenomena, and I examined some of the reasons why
it took so long for scientists to explore them (Kellert 1993, chap. 5).
For various cultural, technological, and scientific reasons, people did
not want to see chaos and in fact were trained not to see it at all. There
was a kind of culturally conditioned resistance to chaotic phenomena.
In reading some of the material about metaphor, we also find a culturally conditioned resistance. The insistence on the literal that one finds
in many self-styled defenders of science and also in some philosophers
comes across as a fear of metaphor—as if metaphor somehow presents
a threat that needs to be either eliminated from respectable knowledge-
seeking discourse or else contained, controlled, and rendered innocuous
by demonstrating its in-principle eliminability.

   For Davidson, the threat of metaphor lies in its open-endedness,
which arises because he sees the interpretation of metaphor as a creative
endeavor ungoverned by rules. He states that "there are no instructions

for devising metaphors; there is no manual for determining what a metaphor 'means' or 'says'; there is no test for metaphor that does not call for taste" (1984, 245). This characterization raises the specter of arbitrary subjectivity as a result of the open-ended quality of metaphorical function: "In fact there is no end to what a metaphor calls to our attention, and much of what we are caused to notice is not propositional in character. When we try to say what a metaphor 'means' we soon realize there is no end to what we want to mention" (262). So we cannot identify its content with what it brings to attention. The "endless" character of potential paraphrases of metaphor leads Davidson to reject the idea of metaphorical meaning. Meaning must be an orderly matter of discrete literal cognitive contents, otherwise it descends into subjective dreamwork.[9]

Rorty, following the work of Davidson, places metaphors outside the realm of the predictable workings of linguistic machinery and considers them thereby unsuitable for theoretical investigation (1991, 166). He contrasts his view with that of Black and Hesse, who see language as a holistic network wherein one change of meaning ramifies throughout the language. In a passage rich with resonances of chaos theory, Rorty writes that

> Davidson's resistance to this "network" view can be put in terms of an analogy with dynamics. In the case of the gravitational effects of the movements of very small and faraway particles (a phenomenon to which Hesse analogizes the insensible but continuous process of meaning-change), physicists must simply disregard insensible perturbations and concentrate on relatively conspicuous and enduring regularities. So it is with the study of language-use. The current limits of those regularities fix the current limits of the cleared area called "meaning." (1991, 164)

Notice that this view implies that the concept of meaning is applicable only in the cleared-out realm of literal language, and metaphor is an unpredictable and hence meaningless stimulus to expand or alter the cleared area. This narrow sense of the cognitive limits it to the content of literal propositions, and with it comes a narrow sense of theory limited to systems of prediction.

In response to the felt threat of such unpredictability, some writers speak of the need to tolerate "chaotic" or "unruly" metaphor in the discovery process (Gentner and Jeziorski 1993, 471, 476), but others insist that their metaphors are in fact to be taken literally. David Porush, for example, states, "When Prigogine and his collaborators describe the traffic jam itself as a dissipative structure, they mean it literally. And I also mean it literally when I describe the novel as fitting the model of dissipative structure. Traffic and fiction are true analogues of each other" (1991, 74).[10] The insistence on the literal, the discomfort with metaphor, reflects a kind of fear of chaos. For metaphor represents a violation of the rules of language, an eruption of the uncontrolled use of words. As we have seen, a number of writers have remarked that the metaphorical use of language is unpredictable and even dangerous. Indeed, a number of rhetorical theorists see figurative expressions as departures from grammatical expectations that introduce ambiguity, uncertainty, and "noise" (Paulson 1991, 43). Yet oddly enough, if we look at metaphor carefully, its breaking of the rules obeys rules of its own. Black actually agrees with Davidson on the open-endedness of metaphorical interpretation and the ambiguity generated by metaphorical rule-breaking, although he does not view these as drawbacks or as barriers to talk of metaphorical content (Black 1993, 24, 29). Similarly, Kittay points out that although the meaning of a metaphor depends on a wider semantic and situational context, it is not for that reason arbitrary (1987, 96–139). Metaphorical categorization brings about a reordering by means of a disordering—a reorganization of conceptual structure. "Metaphor breaks certain rules of language, rules governing the literal and conventional senses of the terms. The rule-breaking takes place not in any arbitrary way but in certain specifiable ways. Hence, we can tell the difference between metaphors and mistakes, and the difference between metaphors and new, technical uses" (Kittay 1987, 24). So metaphor itself is a kind of orderly chaos. As Goodman puts it, "the very novelty and instability that distinguishes metaphor" is "symptomatic neither of uncontrolled caprice nor of impenetrable mystery but of exploration and discovery. I hope that chaos has been reduced, if not to clarity, at least to lesser confusion" (1968, 94).

So not only can we use chaos as a metaphor, but we can think of metaphor as a kind of chaos. And not a threatening, utterly incomprehensible chaos, but a chaos very much like the chaos of chaos theory: unpredictable and yet orderly.[11] Kittay closes her book-length study of metaphor with the following:

> Some order...none the less remains crucial. And a systematic tolerance of a "disorder" that can be shown to be purposeful and intelligible is equally essential, if we are to remain in touch with the well-springs of our creativity and to keep our surroundings

and our minds adaptable to the changing circumstances of our lives and our world. Understanding the workings and the meaning of this latter "disorder" is as much a part of understanding meaning and language as is understanding the "proper order." It is within a carefully conceived "chaos" that metaphors attain an irreducible cognitive content and their special meaning. (1987, 327)

Allow me to spell out the following metaphor: "Metaphor is Chaos." We can understand something new about metaphor if we view it (metaphorically) as the kind of orderly chaos found in nonlinear dynamical systems. The field of language has been structured with a dichotomy between the intelligible and the disordered. Chaos theory offers a more complex differentiation, opening a space for that which is unpredictable and yet patterned, seemingly random and yet rule governed. Seeing metaphor as chaos brings about a conceptual reorganization that invites us neither to celebrate it uncritically nor to condemn it out of hand, but rather to welcome it into our intellectual toolkit with an appropriately circumspect appreciation of its power and its limitations.

6   How to Criticize a Metaphor

My defense of the metaphorical use of scientific concepts should not be read as uncritical acceptance, for just as the persuasive functions of borrowed knowledge can be helpful or inappropriate, its inventive functioning can sometimes go awry. One need not look far for examples of extravagant metaphorical exercises, such when we read in the *National Review* that chaos reconciles science and religion in part because "free will is a strange attractor," an assertion made without argument or elaboration (Mano 1989, 59). On the other hand, Massimo Pigliucci goes too far when he condemns all uses of chaos theory in the humanities as "pseudoscience" because of the appropriators' failure to understand the science, their use of jargon, and their reliance on metaphors that do not improve our understanding (2000, 69). In this chapter I focus on evaluating metaphorical borrowings in order to argue that there are better and worse ways to use scientific knowledge in other fields. I begin by discussing the importance of paying critical attention to metaphorical uses and then develop a set of four criteria for this evaluative task. These criteria include how well the metaphorical use fits the source field, how useful it is for accomplishing its cognitive function, how much it fulfils a genuine need, and how much critical self-awareness the author of the metaphor displays. By providing concrete examples of how to use these criteria to evaluate a number of metaphorical borrowings from chaos theory, I demonstrate that such evaluation is both possible and worthwhile.

Why Criticize Metaphors?

Few have characterized the peril of wayward metaphors as exquisitely as Thomas Reed Powell in his review of the book *The Constitution of the United States: Yesterday, Today—and Tomorrow?*:

> It helps us to know what the Constitution is if we know what it is not. It is a beacon and a Gothic cathedral, but it is not a rock and it is not a beach. Instead of these things, it is a floating dock. . . . It makes you see how marvelous the Supreme Court really is when it can be a balance wheel at the beginning of a chapter and a lighthouse at the end. . . . He brings up in your mind beautiful pictures of the Constitution as a temple and a beacon and a floating dock and he lets you see the Supreme Court shining and balancing in a very wonderful way. I have read a great many books about the Constitution but there is no other book that has given me just the same kind of pleasure that this one has. (As quoted in Ball 1985, 18–19)

Milner S. Ball provides such an extended quotation from this "justly devastating" review in order to make a point that is also a central thesis of this chapter: "metaphors require integrity and the discipline of good writing. Not every metaphor will do" (1985, 19). In his own metaphorical exercise, Ball seeks to challenge the traditional ruling image for the law as "an adamantine structure protecting a space for order and life" and seeks to develop alternative metaphors from the behavior of water. He goes so far as to suggest that we think of the law as plumbing: a structure that provides provisional containment and facilitates the circulation of promises, grievances, and compromises (21, 30). He fully recognizes the danger involved: "As you read, you may come to think that my enterprise is fanciful. You may think that it is fanciful at the very best. You may think that we might as well say that law is a flyswatter or a pasta machine or field peas or a brooding omnipresence. Why not? Could not someone familiar enough with the law rummage around in it and come up with the elements of any image of law she wanted to?" (xiii). But Ball recognizes that he has a responsibility to argue that his proposed metaphor is plausible. In this chapter I focus on just such arguments—on the criticism and defense of metaphors. For, as Ball insists, finding and evaluating new metaphors is not merely a game; new ways of thinking are necessary in order to address the evils done by the legal system (xiv).

Some uses of chaos theory as a metaphor are fruitful and appropriate, but not all of them are. It must be possible to distinguish between helpful metaphorical uses and those that are flaky or pointless or mistaken. Davidson seems to deny the possibility of judging metaphors in this way when he claims that "a metaphor implies a kind and degree of artistic success; there are no unsuccessful metaphors, just as there are no unfunny jokes" (1984, 245).[1] If all metaphors are necessarily successful, then it is impossible to criticize them. But surely, as Max Black has suggested, not just any combination of words can yield an effective metaphor. Consider his example, "A chair is a syllogism"—"In the absence of some specially constructed context, this must surely count as a failed metaphor" (1993, 23). Even Davidson admits that "there are tasteless metaphors, but these are turns that nevertheless have brought something off, even if it were not worth bringing off or could have been brought off better" (1984, 245). Let us leave aside the question of whether a worthless metaphor still qualifies as a metaphor (and whether unfunny jokes nonetheless count as jokes). Instead, let us focus on the question of how to make distinctions between more and less worthwhile, better and worse metaphors.

Metaphors matter, and so the criticism of metaphors matters as well. Just as we can find better and worse metaphors, we can find better and worse ways to critique them. It will not do to simply reject metaphors that strike us as silly, although many of them are in fact ridiculous. Neither can we write them off as merely inspirational gimmicks for the context of discovery, although philosopher of science Alexander Rosenberg seems to do just this. In his discussion of the ways economists have used Darwinian evolution as a metaphor, Rosenberg states, "whether it has been a source of fruitful stimulation is debatable. Whether the theories cut to its pattern have been well confirmed or not seems to me to be the only interesting question for scientists in these disciplines to actually concern themselves with. All the rest is *ad hominem* argument or the genetic fallacy" (1994, 407). Surely it would be fallacious to dismiss a hypothesis simply because its originator misunderstood the source field of the metaphor or because the source field has since come to reject the original theory. But if someone proposes a metaphorical borrowing, an inaccuracy in their characterization of the source field may in fact weaken the analogical "fit" and make the hypothesis less worthy of pursuit. And a previous importation may face challenging potential "antidotes" conceived as analogues to new developments in the original source. My goal here is not to argue whether metaphors "confirm" or "justify" the theories to which they give rise, but to evaluate whether particular metaphors serve useful roles in the overlapping knowledge-creating processes of invention, pursuit, and justification.[2]

Royce de R. Barondes has been one of the most forceful critics of metaphorical borrowing from chaos theory in legal scholarship. Many of his specific criticisms of the work of Scott, Hayes, and others are well taken, but sometimes he faults metaphorical borrowing for not living up to a limited conception of what counts as being scientific. For instance, he takes Scott to task because "the core of Dean Scott's analysis relies on metaphor. The discussion does not reduce the analyzed environment to quantitative terms and apply an analysis to those terms; it does not even specify the variables being considered" (Barondes 1995, 164; see also 171–72). But "metaphorical" should not be used as a term of abuse when rigorous quantitative analysis was never an appropriate goal. The problem here stems from the insistence on applying a positivist conception of scientific methodology to the evaluation of metaphor. If theories are to be judged solely on the quantifiable accuracy of their predictions, then metaphors can be nothing but the sloppiest methodology (Barondes 1995, 173; see also van Peer 1997, 46). But such a conception of knowledge making has been rejected by a consensus of philosophers of science, so it ought not to be used to beat up on people who make use of metaphors. Not even if those people are law professors. The advice that Richard Brown once gave to sociologists holds good for us all: we may feel a need to choose between literalistic positivism and "the airy realms of creative imagination," but this is a false choice. The real choice "is not between scientific rigor as against poetic insight. The choice is rather between more or less fruitful metaphors, and between using metaphors or being their victims" (Brown 1977, 90).

Developing Criteria

Allow me to illustrate briefly the critical appraisal of metaphor by examining a really bad one. As a twelve-year-old student at the Fort Lauderdale Children's Theater, I was given a handout titled "Beginner's Dramatics: Introduction to the Play," which instructed me to think of a play as a pie. The explanation begins by noting that "Many ingredients are needed to cook a pie. . . . Many people and many different kinds of work are involved in a play," and moves on to map the structure of a pie onto that of a play:

> The pie is the *play*. The cook is the *director*. The piecrust—which holds the pie together—is the background for the play, so it must be the *scenery* and the *props*. . . . What kind of pie (play)? Apple pie, chocolate pie, etc.—different kinds of plays. . . . Comedy (ends nicely), Tragedy (ends sadly), Drama (makes you thin[k] seriously). . . . The main ingredient (like apples in pie) determines the kind of play. The apples in the play are the *actors*—the *cast*. (Unattributed handout, 1975)

Notice that whereas the difference in kinds of pie depends on the ingredients, the difference in kinds of plays depends on how a play ends or what effect it has on the audience rather than on which actors are in it. This mismatch in the analogy makes the metaphor a bad fit.

What else is wrong with this metaphor? Consider the choice of pies as the source field. Pies have many different parts, and plays have many different parts as well. But the fact that two things both have many parts does not necessarily mean that one is a good way of thinking about the other. My body, my microwave oven, and my garden all have lots of parts, but each is not necessarily a good metaphor for the others—the relations between the parts are what is crucial. The piecrust holds the pie together; it is in some sense the background for the pie in the same way that the scenery and props are the background for the play. But if "holding together" is the relation between crust and ingredients, does the scenery hold the actors together as the piecrust holds the apples together,? In what sense is the pie plate the stage crew, then? The pie plate seems to hold things together even more than the crust. Ultimately, this metaphor fails to have much cognitive utility for illuminating dramaturgy. But it may serve a different end—the insistent emphasis given the terms suggests that the purpose of the metaphor is not to give us genuine insight into the dramatic arts but simply to teach young people the meanings of certain theatrical terms as well as encourage them to think of their assignment to a stage crew as nonetheless an important job, one that is part of the team.

Although it may be enjoyable to pick apart particular metaphors, my goal in this chapter is to develop a more comprehensive set of tools for their critical evaluation. The scholarly literature on metaphor contains hints about criteria for their evaluation, but no place are the proffered standards collected and systematized. On the basis of existing discussions, I propose the following organization of criteria for evaluating metaphorical borrowings:

1. Fit—how well the elements and relations of the source field can be mapped onto the target.
2. Utility—how well the metaphor performs its structuring or restructuring function.
3. Need—whether the metaphor serves a purpose otherwise unmet.
4. Awareness—whether the metaphor pays explicit attention to its own figurative status.

In the following sections, I explore each of these criteria, explicating them and then illustrating their use in evaluating metaphorical borrowings from chaos theory.[3]

These four criteria do not represent some special set of concerns and considerations that apply only to metaphor. In this I agree with Nelson Goodman, who insists that the metaphorical application of a predicate is fallible and testable, just as is the literal application of that predicate: "we may look again, compare, examine attendant circumstances, watch for corroborating and for conflicting judgments" (1968, 79; see also McCloskey 1985, 80). Perhaps metaphorical predications are less sharp and stable than literal predications, but this is a matter of degree—the testing of metaphors is not radically different from other kinds of testing. On a similar note, Boyd insists that only the diligent criticism of metaphors at the methodological level, rather than linguistic efforts at purification and explication, answers our concerns about the possible vagueness of metaphorical terminology (1993, 523). Susan Haack makes the point well when she says that "metaphor is neither a Good Thing nor a Bad Thing in and of itself; it is, rather, a linguistic device capable of being put to good or bad use" (1998, 72). As one of our tools for making knowledge, metaphor has some distinctive features, but its evaluation need not be so different in the end from the evaluation of any other tool: Does it do what it is supposed to? Does it work better than other tools? Does it do something that we actually want done? It is these sorts of considerations that underlie the four criteria, which I consider to operate like cognitive values, or intellectual virtues that metaphors can exhibit. In what follows, I explore the ways in which these criteria have been employed by some authors in evaluating metaphors and show how they can be used to critique specific examples of metaphorical borrowing from chaos theory.

Fit

Investigators of metaphor have described the importance of a good fit between the source and the target fields in a number of ways. Max Black, for example, urges that the analogies of structure imputed by a metaphor can be examined for "appropriateness, faithfulness, partiality, superficiality, and the like" (1993, 39). Kittay states that in an effective metaphor, "many of the affinities and oppositions are carried along in the transfer of meaning" (1987, 154), and for Gentner and Jeziorski the strength of an analogy rests on "the degree of systematic structural match between the two domains" (1993, 453). Richard Boyd derives this concern for a good fit between source and target from one of the fundamental touchstones of scientific methodology—trying to see whether you might be wrong. He suggests that in evaluating metaphors we need to ask about the reliability, the robustness, and the quality of evidence for the similarities they highlight (1993, 523–24).[4] Metaphorical borrowings across disciplines can suffer from a bad fit due to the special circumstances associated with cross-disciplinary work. Among these potential problems I focus on three and consider them in turn: (1) inaccuracy about the details of the source, (2) excessive flexibility in the use of source terminology, and (3) violations of the theoretical coherence of the source field.

Getting the Source Field Right. Sokal and Bricmont are surely correct when they posit, as one of their general lessons for relationships between disciplines: *"It is a good idea to know what one is talking about"* (1998, 185, emphasis in original). But accusing someone else of making mistakes about the science is a tricky game. Mirowski's book *More Heat Than Light* (1989) details how the founders of neoclassical economics misunderstood basic physics. But Hal Varian (1991) has turned the tables and accused Mirowski of misunderstanding neoclassical economics, specifically the integrability conditions for prices to be treated as a vector field. D. Wade Hands (1993, 129) agrees that Mirowski makes mathematical mistakes, but he says that the mistakes can easily be taken into account and thus do not weaken the overall argument. It is worth noting that sometimes even scientists get their science wrong; the criteria for good cross-disciplinary metaphorical scholarship sometimes also apply to the most literal-minded scholarship in a single discipline. So in evaluating metaphorical borrowing, as in evaluating any argument, it is not enough simply to catch someone making a mistake—we must ask: What is the import of that mistake? Sometimes an erroneous understanding of the source field can render the metaphorical fit so weak as to be useless, whereas in other situations the error concerns an aspect of the source field that never needed to be carried over in the first place. So mistakes about the source field need not render the metaphor worthless; "Man is a Wolf to Man" successfully conveys a sense of human cruelty, despite the fact that wolves are not actually malicious.

Indeed, historians of science have shown how misunderstandings can turn out to be downright helpful. I. B. Cohen recognizes that although borrowings from the natural sciences sometimes suffer from "incorrect science," such errors might not adversely affect claims of only a general sort of similarity (1993, 31). Indeed, Cohen states that many fruitful advances in social thought have come from situations in which the concept originating in natural science was not fully understood; Darwin got Malthus wrong, and Freud misread Helmholz (1994, 69).[5] And whereas Mirowski may be correct that neoclassical value theory suffered from a misappropriation of physics, Avi Cohen insists that neoclassical economics nonetheless gives theoretically coherent and empirically supported accounts of some aspects of prices (1993, 216).[6] We may agree that it is a good idea to know what one is talking about but at the same time concur with Ivor Grattan-Guinness that "the incompetent transfer of theories from one science to another does not forbid fruitful results from being obtained" (1994, 105; see also Shusterman 1998, 128–29).

Returning to the example of Robert E. Scott, we find he suggests that "we should look to Chaos Theory as a metaphor for the way to think about the contradictions and the tensions inherent in the legal system. At the initial level, Chaos Theory tells us to accept the contradictions, the disorder, in our legal system. All systems, including the legal system, are unpredictable and erratic" (1993, 348). However one feels about the advisability of accepting disorder, Scott's argument suffers because he bases it on a mistaken premise. He claims that the first lesson of dynamics is that

> all systems are chaotic, in the sense that they are subject to irregularities that make predictions of outcomes in particular cases impossible.... The butterfly effect teaches us that small differences in initial variables will always produce dramatic variations in final outcomes. By explicitly applying this to law, it becomes clear that even slight differences in the facts of cases result in

wildly disparate judicial outcomes. In both instances, disorder
is inevitable. (1993, 348)

But it is simply not true that all physical systems are subject to the
butterfly effect (clocks, for example); despite the enthusiasm of some
popularizers, sensitivity to initial conditions is merely widespread, not
universal. Furthermore, as Royce de R. Barondes has pointed out, Scott
mistakes the indeterminism of legal decision-making for sensitivity to
initial conditions: the fact that judges may issue different results in
identical circumstances actually presents a strong disanalogy with the
strict determinism of chaotic systems (Barondes 1995, 172).[7]

Unfortunately, Scott is not alone in extrapolating from sweeping
claims that chaos theory has never made. When Gordon Slethaug claims
that "the dark and tragic side of the complex relationship inherent in the
stochastic process is exemplified in the life and death of Paul [in Norman
MacLean's *A River Runs Through It*], for, as chaos theory suggests,
order and stability must finally give way to turbulence and randomness," his borrowing suffers from the fact that chaos theory suggests no
such thing (2000, 70). Neither does it suggest that order and randomness must alternate (Demastes 1998, 83), or that randomness must give
way to order, for that matter. Another common mistake that fatally af-
flicts borrowings from chaos theory is the notion that strange attractors
are mysterious entities exerting physical influence on actual systems. For
example, much of the last chapter of Harriet Hawkins's book *Strange
Attractors* is taken up with the extended metaphor of Shakespeare's
Cleopatra as a strange attractor. Hawkins holds that, like Cleopatra,
"the strange attractor can also be seen as an outside force" that inter-
venes to instigate chaos (1995, 126). But strange attractors are neither
forces nor external to the systems they characterize. A related error occurs in the work of legal scholar J. B. Ruhl, who conceives of freedoms,
rights, and regulations as three attractors vying for dominance in the
interactions between law and society. For Ruhl, approaches such as allowing citizen suits to enforce regulations constitute a hybrid of two or
more approaches (1996, 855, 874). But although a system may have
multiple attractors, each surrounded by a "basin" of points that eventually converge to it, these attractors coexist without interacting and
cannot combine into new intermediate forms.[8]

Examples such as these could be multiplied until readers lose both
their interest and their tempers. I believe I have established that, on some
occasions, getting the science wrong dooms an attempted metaphorical

borrowing. Let us consider the converse point, however: not all mistakes are fatal. For example, when Thomas Weissert misreads the bifurcation diagram of the logistic map as representing paths with choices, his analogy with the "forking paths" in Borges's garden fails (1991, 236–38). But the analogy would work fine if he substituted the very different bifurcation diagrams found in Prigogine's work, which do in fact depict alternatives selected by a system as it evolves.

Sometimes errors represent nothing more than honest and harmless mistakes. Adelaide Morris says an attractor is a point, which is strictly false, but she then goes on to speak of limit cycles and other nonpoint attractors, so her "error" should not be used to discredit her main contention (1991, 212). Even the economists William Baumol and Jess Benhabib, in their survey article on the significance of chaos theory for economics, mistakenly describe the rectangle on the cobweb diagram as the attractor, when the attractor is actually just the two points to which the orbits converge (1989, 91). On other occasions, a mistake may be more serious but still not impugn the author's conclusion. Peter Mackey does a fine job describing chaos theory, yet makes the unfortunate claim that "in an unpredictable system . . . no causal links join an occurrence at one time and an occurrence at another" (1999, 39). Still, this denial of causality plays no argumentative role in what follows. Similarly, although Michael Wise makes some minor errors in his presentation of nonlinear dynamics, the mistakes never play any significant role in his later argument. Thus, he is correct for at least some instances of borrowing, "technical details are less important than the philosophical and cultural implications of changing expectations about what science can do" (1995, 752).

The concept of sensitive dependence on initial conditions—the so-called butterfly effect—has proven to be one of the most popular candidates for metaphorical borrowing, and it provides a valuable case for examining how well such metaphors fit. In the field of literature, for example, John Barth observes that many conventional novels have butterfly effects of disproportionately great consequences for minor events, including Thomas Rogers's *The Pursuit of Happiness* and Kurt Vonnegut's *Deadeye Dick* (1995, 331).[9] Some of the most extended discussion of sensitive dependence in literature has concerned James Joyce's novel *Ulysses*, which tells us the external details and inner life of one Leopold Bloom and some of his close associates as he wanders about Dublin on one day in June. In the "Aeolus" chapter, one character reflects, "I have often thought since on looking back over that strange time that it was that small act, trivial in itself, that striking of that match, that determined the whole aftercourse of both our lives" (Joyce 1986, 115, chap. 7, ll. 763–65). For Peter Mackey, this passage provides a clue to the author's method, especially when considering that Joyce is said to have remarked, while writing *Dubliners*: "it is my idea of the significance of trivial things that I want to give the two or three unfortunate wretches who may eventually read me" (Mackey 1999, 28). Mackey details how such elements as "attention to trivial things, contingent relations, considerations of chance and determinism, and underlying schemas or ordering devices amid seeming chaos" provide convincing links to chaos theory (35). And Rice draws the conclusion that because lives and chaotic systems share sensitive dependence, they "allow any triviality to make all the difference. Every instant becomes ripe with possibilities that can work for or against us" (1997, 157). Notice that here Rice attends to the fact that sensitive dependence applies to each and every trivial event, as opposed to the isolated critical points associated with specific moments of unstable equilibrium.

In the field of law, metaphors of sensitive dependence have raised more spirited debate. Some, such as J. B. Ruhl, find analogies to the butterfly effect in the evolution of regulatory regimes: "If, at certain crucial turning points, the system's trajectory had been slightly perturbed, the system's trajectory may have traversed a different path through phase space," and there would be no Environmental Protection Agency today (1996, 866; see also Shaviro 1990, 100). Other authors find sensitive dependence in the process of legal decision-making, suggesting that just like the butterfly, "so, too, a defendant's inappropriate smile in the course of a criminal trial might generate the cognitive armature around which jurors weave a story of guilt" (Sherwin 1994, 695). Reynolds suggests that each Supreme Court judgment can have large effects on the disposition of new cases in terms of decisions as to whether to settle, whether to file suit, and how later cases are argued (1991, 114).[10]

Not everyone, however, is convinced that legal systems behave in a way that corresponds to sensitive dependence. Susan Bandes accepts that chaos theory has some valuable lessons for the law but insists that the analogy to sensitive dependence assumes "a stable environment, a jigsaw puzzle with one piece removed" (1999, 497). For Bandes, the system of common law is much more complicated and organically interconnected than a controlled physics experiment: "no discrete systems at work here"—just a welter of interacting forces and effects, so we cannot predict what would have happened had a small change occurred (497). M. B. W. Sinclair, alarmed by postmodernist legal theorists' attacks on the notion of legislative intent, seeks to disarm the notion of sensitive dependence in law, insisting that there are good reasons for law to not be too sensitive to facts. For Sinclair, the law needs to be rough and general enough for us to obey it without minute attention to details of a situation. Whereas any fact might be significant, sensitive dependence would require that all facts are; but surely some facts truly are insignificant for a given case (1997, 373). Sinclair goes on to argue that law is not sensitively dependent on the conditions of its creation either, because judges' interpretive power can steer the law in new directions regardless of where it started (373). It is not my goal to evaluate these competing arguments and reach a conclusion about whether legal systems exhibit something analogous to the butterfly effect. My point is just that it is possible to engage in critical evaluation of metaphorical claims, in this case by posing analogies and challenging them on how well they fit.

Flexible Terminology. Flexibility of terminology represents a second dimension of fit: on some occasions the meanings of words from the source field may be stretched to the breaking point, rendering the metaphor weak or even ridiculous. Hayles raises the issue aptly in her introduction to a collection of essays on chaos theory and literature: "Often the debate comes down to questions of language. Should terms appropriated from chaos theory be confined to their technical denotation, or is it valid to use them metaphorically or analogically? If they are used metaphorically, what do such arguments demonstrate?" (1991, 15). Similarly, Ronald Shusterman suggests that recent conflicts between the humanities and the sciences call for answers to questions such as "When is a metaphor permissible? When is it useful?" He suggests that progress can be made in answering these questions by "deciding which language-games can tolerate violent transfer of symbolic schemata and which demand more rigorous vocabularies" (1998, 131). In contrast, I believe that criteria for the usefulness of metaphors should be sought through the examination of specific instances rather than at the general level of entire disciplines. And laying down conditions for the permissibility of metaphors is no part of my project.

Shusterman is responding to the hostility displayed by some self-styled defenders of natural science against borrowings by those in the social sciences and humanities. This hostility sometimes takes the form of attacks on metaphorical flexibility. For instance, Frank Miele of *Skeptic* magazine complains that the worst abuses of chaos theory "occur when terms that have a clearly defined meaning, usually mathematical, in the physical sciences are imported into literary studies as metaphors." He uses as an example the notion of an attractor, which "really makes sense within the world of physics and non-linear mathematics...but becomes meaningless chaobabble when applied elsewhere" (2000, 54).

Another strong warning against metaphorical flexibility comes from the mathematician René Thom. In an essay decrying what he saw as the fashionable championing of randomness, he writes (in a section with the acid title "On Fuzziness in Concepts or Deontology of Vulgarization") that concepts such as determinism, chance, and randomness "acquire a precise signification only in the framework of an explicit (mathematical) formalism. By not attaching these concepts to the formal framework which renders them precise, one is condemned to hold discourses not necessarily senseless, certainly, but of a fluidity, an ambiguity such that nearly always they spill over into verbalism" (1983a, 78).[11] Sokal and Bricmont also warn that "because mathematical concepts have precise meanings, mathematics is useful primarily when applied to fields in which the concepts likewise have more-or-less precise meanings" (1998, 9). They urge that the natural sciences not be seen as merely a source of metaphor, because in the context of science, terms have specific meanings that differ from their everyday ones: "If one uses them only as metaphors, one is easily led to nonsensical conclusions" (187).

Sometimes the problem of flexibility arises because terms are treated as meaning the same thing when they do not. And sometimes the problem arises because, as Argyros puts it, metaphorical similarities "can tend toward such generality that they end up saying practically nothing" (1991, 227). Playwright Michael Frayn, who borrowed from quantum mechanics to explore the uncertainties of human motivation in his play *Copenhagen*, writes in the postscript to that play that the concept of uncertainty "is one of those scientific notions that has become common coinage, and generalized to the point of losing much of its original meaning" (1998, 82). He resists such excess of generality by attention to the details of the scientific concept and by taking pains to highlight the differences between the technical scientific usage and the extended analogy that the play draws with human thoughts and intentions.

In recognizing the limits of his analogy, Frayn displays the virtue of metaphorical awareness that I discuss later in the chapter. And I also consider many specific examples of the vices of equivocation and overgeneralization. But is scientific terminology really so precise that it cannot abide some flexibility? Mirowski points out that the "laxity" in the term "force," which was "used to refer to diverse and sometimes contradictory notions," was helpful in combining "a formidable set of connotations and concerns under one umbrella word" and probably helped encourage the widespread simultaneous discoveries of the concept of energy conservation (1989, 47). Indeed, different economists arrived at very different theories while appealing to the same physical metaphors, demonstrating "the free play inherent in metaphorical reasoning" (170). The interaction view seems to presuppose two fields that are initially fairly stable, whereas metaphor can unsettle things and open a space for new categorizations that might stay open indefinitely. Evelyn Fox Keller has examined the borrowing from one science by another, for example, from cybernetics to developmental biology, and she has argued that the explanatory power of these metaphorical borrowings depends crucially on their lingering ambiguity (2002). When does the flexibility of a scientific term facilitate equivocation and sloppy generality, and when does it serve as a useful resource for conceptual inventiveness? A pluralistic attitude toward meaning would encourage us to examine each case separately, without insisting that the univocal meanings appropriate for some mathematical pursuits are the only kinds of meanings worth having. Notice that I say that precision is appropriate for some mathematical pursuits because the philosopher of science Bas van Fraasen (among others) has said that flexible meanings are not necessarily defects even in mathematics and natural science: "Quite to the contrary, ambiguity and vagueness not only characterize also our most precise discourse at every stage of our history but are essential to the character of discourse" (2002, 145).

The problems associated with flexible terminology certainly plague borrowings from chaos theory. After all, a term such as "chaos," so evocative of everyday connotations, makes for a situation especially ripe for portentous equivocation.[12] Certainly the "evocative and polysemic richness" of the term "chaos" accounts for some of its popularity among scientists as well as the general public (Mackey 1999, 36; see also Hayles 1990, 22). But as Matheson and Kirchhoff point out, "one should beware of the assumption that every occurrence of the words 'disorder' or 'chaos' in any discipline . . . is necessarily related to chaos theory" (1997, 32). Philip Mirowski, while acknowledging the usefulness of flexible terminology in some circumstances, comes down hard on those who make use of chaos theory in economics: "The terminology of 'chaos' has already been sadly abused in the literature, since it should only legitimately be attached to dissipative systems of flows (and then, not to all of them)" (1990, 301–2). As a matter of fact, a great deal of work in physics has been done on chaotic behavior in nondissipative (or "Hamiltonian") systems, so Mirowski seems a bit overeager to police the use of the term here.[13]

Of course, scientific terms such as "force" or "energy" have a vague meaning when used in nonscientific contexts. But terms from nonlinear dynamics sometimes have their meanings stretched to the breaking point. Consider, for example, the suggestion that the verse "Humpty-Dumpty" exhibits fractal self-similarity because it rhymes (cited in Eoyang 1989). Hayles provides a valuable account of how the term "linear" has come to have connotations beyond the technical meaning: especially due to Michel Serres, this term "became associated with a whole set of values in critical discourse, ranging from binary logic to abstraction and rationality" (Hayles 1991, 18). In this way, anything considered admirably paradoxical or circuitous can be labeled nonlinear and trumpeted as having anticipated the science of chaos.[14] Gordon Slethaug performs a similar act of stretching on the term "recursive," until it reaches a point at which he admits that "because of the necessity for creativity in art along with the recognition of patterns, *everything*, I am tempted to say, is both recursive and iterative, following familiar patterns of plotting but trying to incorporate individual differences" (2000, 98; see also 78). Surely a term that applies to everything has lost its usefulness.

As we saw in chapter 3 on the rhetorical functions of borrowing, terminology sometimes is appropriated in order to convey the flavor or the prestige of the source field. Thus, every recurrence becomes "recursive," every repetition becomes an "iteration," and every similarity becomes a "self-similarity." Thomas Rice makes such free use of the term "sensitive dependence" in his discussion of Joyce that nothing ever seems simply to depend on anything else—it "sensitively depends" on it (1997, 95, 97, 98). On occasion, the flexible use of scientific terminology rises (or descends) to the level of the pun, as when Rice describes the way the meaning of the word "period" becomes ambiguous in Joyce's *Ulysses*. At some points in the text, "period" means the punctuation mark, whereas at other points it refers to the menstrual cycle. When Rice observes that this is "literally a case of 'period doubling'" (1997, 99), it is difficult to know what sense of "literally" is at work. Although we should reject any strict separation between the persuasive and the inventional functions of metaphor, uses such as these are best understood in terms of stylistic effect rather than conceptual reordering.

Retaining Coherence. Attentiveness to potential dissimilarities between source and target is admirable. Such mismatches may arise in any metaphor, not just metaphorical borrowing across disciplines. They may stem from important differences between the particular elements of the two fields, or from differences in the relationships between their respective elements. As an example of mismatch between elements, Lawrence Cunningham suggests one reason for questioning the analogy between the fluctuation of securities prices and the random Brownian motion of microscopic particles: people are sentient, and prices depend on their awareness, whereas molecules have no such awareness (1994, 570). In this case, as with other instances of criticism, defenders of a metaphor may respond that the dissimilarity between the elements poses no problem for the usefulness of the analogical model. In other cases of borrowing, the alleged mismatch arises not from a difference between the elements but from an incoherent translation of the logical structure of the source field.

Mackey voices a concern for this aspect of fit when he states that "it seems fair to ask that even metaphoric uses of another discipline's ideas employ the logic of those ideas soundly" (1999, 14). Another example of this concern comes from George Gumerman and David Phillips Jr., who criticize their fellow anthropologists for borrowing the concept of the "ecotone" from ecology. They argue that their colleagues have been too hasty in translating from the notion of a transition zone between biological regions to the notion of transition zones between cultures: "Archaeologists cannot simply adopt models because they are useful ways to organize particular sets of data; a more adequate justification for the process of selection is needed, one that examines the basic validity of those models in their new application. To do so it is probably necessary to consider not just the particular model used to organize data, but its entire logical development within its own body of theory" (Gumerman and Philips 1978, 187). In these calls for attention to the logical context of borrowed ideas, we may discern a criterion for evaluating metaphors based on their coherence.

One of the greatest champions of the importance of coherence is Philip Mirowski. Indeed, his book on economists' metaphorical borrowing from physics has been described as angrily using "a battering ram on behalf of strict theoretical coherence" (de Marchi 1993, 3). One of Mirowski's central points is that the economists who founded neoclassical theory borrowed the mathematical structure of vector fields from physics but never included anything corresponding to the conservation of energy: "this suppressed conservation principle, forgetting the conservation of energy while simultaneously appealing to the metaphor of energy, is the Achilles heel of all neoclassical economic theory, the point at which the physical analogy breaks down irreparably" (1989,

231). No doubt there is a mismatch, but Mirowski acknowledges that others may differ as to whether the disanalogy causes irreparable harm and renders the appropriation from physics worthless: "it is certainly true that one need not be obsessed with the exact duplication of all aspects of a metaphor when it is transported from one area of inquiry to another. However, one of the most attractive aspects of analogical reasoning is the prefabricated nature of an interlocked set of explanatory structures and constructs, allowing quickened evaluation of logical coherence" (272). He insists that one cannot simply pick and choose which aspects of the metaphor to keep: a concept, like that of a vector field, "possess[es] a modicum of structural regularity that, if absent, undermines its logical integrity" (272).

For Mirowski, "the trick to metaphorical evaluation is an ability to sense when one has finally ventured beyond the pale, so that the coherence of the metaphor is strained to the point of dismemberment" (1989, 314). In some cases, the distortions of logical structure may be fruitful, but in other cases they undermine the very intellectual credibility of the enterprise or "destroy the internal coherence of the metaphor and violate its mathematical consistency, rendering it useless as a paradigm of research" (301; see also 372). Although logical coherence proves a useful tool in the evaluation of metaphorical borrowing from chaos theory, I do not venture to judge whether Mirowski has been excessive in his employment of it. Sometimes a battering ram is just the right tool for the job.[15]

With regard to borrowings from chaos theory, we find a mismatch between the logical structures of the source and the target field in Slethaug's discussion of the blending or admixture of different genres in contemporary fiction (2000, 135–40). He refers to this as their stretching and folding, but whereas blending genres may correspond to "folding," nothing is proffered to match the equally crucial "stretching."[16] This example of metaphorical borrowing may not achieve the level of incoherence that Mirowski attributes to neoclassical economics, but it shows a marked lack of fit with the mathematical structure of stretching and folding so central to chaos. We find a more modest failure of theoretical coherence in Jason Scott Johnston's essay on the chaotic dynamics of legal change. Johnston examines the way different "regimes" of judicial decision-making can lead to oscillations in the way judges prefer to operate:

> If rules pull toward balancing and balancing toward rules, it follows that the formal structure of the law may undergo self-sustaining and potentially endless cyclical changes.... Judges

may begin with rules, and then adopt balancing, and revert to rules, and so on. Moreover, the frequency of change may increase with the volume of cases. . . . Finally, it may be impossible to predict changes in form. Form depends in part on the path of litigated cases, which in turn depends on the random circumstances which generate cases. . . . Thus, even if the legal process was not itself random, legal form would be. On this theory, the path of the law is hardly a "path" at all: it is a series of nonlinear, chaotic jumps from point to point. (1991, 362)

Unfortunately, Johnston relies on the unpredictability of external inputs for the "chaotic" behavior, meaning that what he describes is not endogenous randomness at all. The most convincing elements of the analogy argue only for some kind of oscillatory behavior, which could be fleshed out with some quantitative analysis of numbers of cases and decisions. Without any argument that the model could show endogenous randomness, there is as yet no reason to speak of chaos.

Utility

Recall how Lawrence Tribe framed his borrowing from physics in terms of the power of structuring metaphors to generate potentially valuable hypotheses: "my criterion of appraisal is whether the concepts we might draw from physics promote illuminating questions and directions" (1989, 2). Such an inventional use represents one of the possible structuring or restructuring functions that metaphor can serve, and utility in serving some such function is one criterion for evaluating metaphors. Nelson Goodman adumbrates this criterion by saying that when reorganization results from metaphorical transfer, "new associations and discriminations are also made within the realm of transfer; and the metaphor is the more telling as these are the more intriguing and significant" (1968, 79). Analogies that lack such utility are characterized by I. B. Cohen as "inappropriate," meaning that they fail to advance the subject at hand (1993, 32).

One aspect of utility is productivity, which turns on how many possible extensions and ramifications the metaphor opens up. Some metaphors make their point and it ends there. At other times, a rich stock of further implications remains to be elaborated, and Max Black holds that such productive resonance can "strengthen a metaphor and heighten its interest" (1993, 40; see also Mirowski 1989, 278). Eva Feder Kittay draws our attention to Plato's metaphor of teaching as midwifery, which is powerful because it evokes a complex and varied set of relations that can be extended into contiguous fields such as mating and childbirth. Kittay calls "the ability of a metaphor to be utilized for the expression of new but related ideas" the *productivity* of a metaphor—the capacity to spin out new implications that we associate with good metaphors (1987, 286; see also Brown 1977, 124; Klamer and Leonard 1994, 36–37). One dimension of productivity is the richness of the structure that the source field has to offer. Lakoff and Johnson mention two metaphors for the human mind: mind-is-machine versus mind-is-brittle-object. The first gives us several things to focus on (on/off, efficiency, output, mechanism, energy source), but the second "is not nearly as rich. It allows us to talk only about psychological strength" (1980, 143). There is a trade-off here with a competing dimension of productivity that Richard Brown calls "economy" or "ease of representation and manipulation, both conceptually and practically" (1977, 104). Rich and elaborate concepts may be difficult to use, whereas a concise, economical metaphor may be easier to work with but less conceptually rich.

Just as scientific theories may be judged on their fruitfulness for future elaboration as well as on the results they have already generated, metaphorical borrowings should also be fruitful. For example, I. B. Cohen notes that Newton's mechanics by itself never provided a useful model for the social sciences because it was too abstract to give a visualizable, mechanical image (1993, 33). In connection with the importance of productivity, Cohen also cites Ménard to the effect that for a conceptual transfer to be fertile, it must allow the original idea to be decentralized so as to keep possible an awareness of the important differences as well as similarities (Cohen 1994, 68). It is just this kind of awareness that serves as our fourth criterion of good metaphors.

Kittay points out that after a lot of the work is done, we can summarize the analogy between two fields in a nice diagram. "But such a diagram does not do justice to the stretching and pulling required in order to match elements in the two fields. If the match is too easy, the metaphor is banal, and its effect is decorative rather than cognitive: since the use of the vehicle field does little to restructure the topic field, the use of a mediating metaphor is relatively gratuitous" (1987, 287). Shelly's metaphor of workers as bees is "no doubt effective rhetorically," but "not all metaphors have much cognitive interest. The degree to which a metaphor is enlightening depends rather on the degree to which the vehicle field is going to be productive of new meaning and new insights into the topic domain" (288).

Notice that an exclusive insistence on a good fit between source and target field renders all metaphors suspect. After all, the only way for the two fields to match perfectly is for them to be either exactly the same field or two fields with identical structures, and in either case the metaphor would hold little conceptual interest. Again we see that a metaphor is more than merely a comparison or a pointing out of a preexisting similarity: a genuinely interesting metaphor involves the interaction of two fields with important differences in their structure (Kittay 1987, 29). So the virtues of fit and utility must be balanced. Goodness of fit between source and target does not guarantee a good metaphor: the fit may be so exact as to make the metaphor redundant. Conversely, a lack of fit between source and target does not necessarily make the metaphor bad: some degree of mismatch is required to make the structuring or restructuring work. The presence of multiple competing demands makes it both easy and hard to criticize metaphors; it is easy because no metaphor can fully satisfy all the demands at once, and it is hard because evaluating metaphors requires the use of judgment to determine whether the tradeoff between demands has been accomplished appropriately. In this way, the evaluation of metaphors resembles Thomas Kuhn's description of the evaluation of competing scientific theories. The operation of multiple and competing intellectual values makes evaluation of theories difficult but not merely subjective (Kuhn 1977).

Adelaide Morris provides an apt example of the usefulness of chaos theory for bringing structure to a target field—in this case, the work of American poet H.D. Morris's thesis is that "the coincidence between H.D.'s mythopoeic mind and the science of chaos offers rich and resonant access to aspects of her work that have been difficult to capture through conventional literary analysis" (1991, 213). Some of these aspects are indeed striking, as when she characterizes the long poem *Trilogy* as the poet's search through war-torn London "for the pattern that pulls all else toward it, a pattern that prevails across scales, through time and over space, a pattern whose every recurrence mixes characteristics that are 'the same—different—the same attributes, / different yet the same as before'" (213). Here the metaphor of the strange attractor enables Morris to conceptualize and analyze a work of literature in new and interesting ways. In contrast, Gordon Slethaug's marshaling of chaos theory to discuss Norman MacLean's *A River Runs Through It* seems to lack any substantive structuring function. Certainly there is extended metaphorical imagery connecting the River and Life. But little is added by imposing talk about chaos onto MacLean's descriptions of turbulence. Both MacLean and chaos theorists talk about fluid flow, but no new insights arise from the comparison. For example, nothing is gained by noting that the character Paul casts figure-eight loops while fishing and the Lorenz attractor looks a little bit like a figure eight when you lean it on its side and squint (2000, 71). Note also that many metaphorical borrowings from chaos theory make little use of the richness of nonlinear dynamics, indicating a lack of productivity.

Metaphorical borrowing can be useful for restructuring as well, as we may note in the utility of chaos theory as both an antidote for previous importations from physics and as a tool for challenging unquestioned dichotomies. A recent article cowritten by professors of law, biology, and physics provides an example of how chaos theory can serve as a metaphorical antidote. Edward Adams, Gordon Brumwell, and James Glazier explore the implications of nonlinear dynamics for the *Palsgraf* case that has served a crucial role in tort law (*Palsgraf v. Long Island R.R. Co.*, 162 N.E. 99 [N.Y. 1928]). The case concerned a Mrs. Palsgraf, who was injured while standing on a train platform. Another passenger was being pulled from a moving train, and when his package of fireworks exploded the sound vibrations dislodged a ceiling tile that hit Mrs. Palsgraf. She successfully sued the railway, but the decision was reversed by the New York Court of Appeals, which held that the actions of the railway company were not the proximate cause of the injury. To prove the tort of negligence, one must establish a causal link between the defendant's breach of a duty and the damage to the plaintiff. Proximate cause, still a widely used notion, implies a limitation to "situations where risks are foreseeable as likely to occur because of a negligent act" (Adams, Brumwell, and Glazier 1998, 511). The causal linkage cannot be too long or tenuous—it needs to be a simple chain from one variable to the next. Thus, the notion of proximate cause is based on the linear dynamics of clear proportionality between cause and effect, and relies on a notion of foreseeability that presumes linear predictability. It betrays a "linear bias" (514).

Adams, Brumwell, and Glazier characterize the *Palsgraf* situation as chaotic because a small change could have made a big difference, the behavior of the tile was strongly nonlinear, and it was entirely governed by deterministic equations. Of course, they do not attempt to actually reconstruct a strange attractor for the behavior of the railway station because there are far too many variables involved, and the parameters change so rapidly (1998, 525, 542). Instead, they rely on insights from chaos theory to draw attention to certain aspects of the *Palsgraf* situation that are analogous to chaotic behavior. In the face of such nonlinear and unpredictable behaviors, they argue, legal thinking should replace the linear notion of proximate cause with a broader conception of liability. They suggest that the railway's responsibility was to maintain the system within a safe range of perturbations, by testing and correcting for small perturbations and working to ensure that larger, untested perturbations do not occur. Thus, one can be liable even for outcomes one cannot have foreseen, and this expanded notion of duty flows from an understanding of the complicated causal relationships that can affect chaotic systems (539). In this way, the application of chaos theory leads to a replacement of the linear notion of causation that had previously been enshrined within legal thinking.[17]

Chaos theory can be useful for challenging a number of presumed dichotomies, as when Adelaide Morris observes that "what chaos theory gives us is a way to think of order as a delicate interplay between the forces of stability and the forces of instability, between the pull toward similarity and the generation of difference, between determinism and unpredictability" (1991, 215; see also Hawkins 1995, 151; Rice 1997, 10). Of course, not every reconciliation of opposites requires reference to chaos theory, as those familiar with the work of Hegel will insist. Nonetheless, one especially apt use of chaos theory to restructure an existing field concerns the notion of personal identity. In discussing the character of Leopold Bloom in *Ulysses*, Mackey uses chaos theory to argue against those who say that Bloom's identity is a formless free association with nothing to determine its wandering (1999, 124). On the contrary, Bloom's personality is not just a mess, or the sum of overlapping external influences—chaos theory can help us make sense of a notion of identity beyond both strict predictability and utter randomness. Although Bloom's consciousness is ever changing, recurring desires and concerns help to define an abiding identity "as an eddy abides its shape" (128; see also Oliver 1991). According to Ira Livingston, this use of chaotic patterns as an image of modern individuality bears "the signature of a Romantic ideology" about "mobility and transformation" and "identity based in a kind of change" (1997, 11). Livingston is especially concerned to highlight the ways that Romantic poetry insists on the social and economic conditions of its possibility. Romantic poets accomplished this in part by taking upon themselves the mantle of "dysrhythmia," the irregular rhythm of the professional poet as distinguished both from the regular order and the disruptive disorder of agricultural or factory work. Not bound to the patterns of the sun or the seasons or the clock, nor subject to the randomness of natural disasters and economic dislocations, the literary professional can chart his or her own individual pattern (159). By providing an alternative to both the regularity of the clock and the hurly-burly of the pressure cooker, chaos theory proves its utility for restructuring an existing field in a way that rejects alternatives thought to be mutually exclusive.

Need

Fit and utility do not always suffice to make a metaphor worthwhile; it should also serve some new and distinctive need. One dimension of whether a metaphor is needed can be labeled novelty, the degree to which the metaphor contributes new perspectives on the target. Nelson Goodman champions the importance of novelty in declaring that "as there are irrelevant, tepid, and trivial literal truths, there are farfetched, feeble, and moribund metaphors. Metaphorical force requires a combination of novelty with fitness, of the odd with the obvious. The good metaphor satisfies while it startles . . . metaphor is most potent when the transferred schema effects a new and notable organization rather than a mere relabeling of an old one" (1968, 79). A second, related dimension of this criterion is distinctiveness: the degree to which the source or focus is especially suitable and others are lacking. Max Black suggests that we examine metaphors in terms of the possibility of substitute terms: "metaphors that survive such critical examination can properly be held to convey, in *indispensable* fashion, insight into the systems to which they refer" (1993, 39, emphasis added; see also McCloskey 1985, 78).

Matheson and Kirchhoff present a scheme for ranking the strength of analogies between two fields, based on how distinctively necessary the source field B is for understanding the target field A. In analogies of the first type, "B allows us to learn about A in ways that could not be accomplished without modeling A as B"; for example, some acoustic problems are only solvable by using analogous electrical circuits (1997, 41). In analogies of the second type, "those unacquainted with A but knowledgeable about B can best learn about A by pursuing A's analogy with B"; for example, learning about gases may best be accomplished using knowledge about billiard balls (41). The third type of analogy involves situations in which "those who are somewhat acquainted with A can learn more about A by coming to learn about B," and the final category consists of situations in which "B is structurally similar to A, and this is inherently interesting" (41). We need not accept distinctiveness as the sole measure of an analogy's strength or of a metaphor's worth, but it can surely play a role in evaluation.

An instance of borrowing may fail to exhibit the virtue of distinctiveness when golden opportunities are missed. For instance, in his discussion of Jorge Luis Borges's story "The Garden of Forking Paths,"

Thomas Weissert repeats the common claim that relativity, quantum mechanics, and chaos theory are the three great revolutions leading to postmodern science. He states that "the second revolution, quantum mechanics, is not discussed in this paper because relativity theory and chaos theory are more appropriate to Borges's work" (1991, 241). This judgment of appropriateness will strike those familiar with alternative interpretations of quantum theory as painfully unfortunate, because the "many-worlds interpretation" posits branching structures of possible universes that resonate so strongly with this story. The omission of this strikingly parallel conception of quantum mechanics from his discussion makes Weissert's appeal to chaos theory seem especially forced and unnecessary.

Does chaos theory really offer anything new and different? Especially with regard to the notion of sensitive dependence in the social sciences, we may be tempted to agree with Sokal and Bricmont that "human societies are complicated systems involving a vast number of variables, for which one is unable (at least at present) to write down any sensible equations. To speak of chaos for these systems does not take us much further than the intuition already contained in the popular wisdom" of horseshoe nails or Cleopatra's nose (1998, 146). The question of how far a particular metaphorical borrowing takes us, however, might reasonably be settled by examination of each case rather than a sweeping dismissal. Some cases readily yield a judgment that chaos theory offers persuasive power rather than inventional utility. For example, one of Harriet Hawkins's points is that "in literature, as in life, momentous, tragic and unforeseeable results often come from very small causes," and there is hardly much novelty in asserting this (1995, xi). Thomas Rice notes that finding analogues to the butterfly effect in everyday life is a commonplace observation, and that the passage in *Ulysses* about the momentous consequence of the striking of a match discussed above is "a portentous narrative cliché, its banality reinforced by its awkward style. (The second sentence features six *thats*)" (1997, 82, 95). So metaphors with adequate fit and reasonable utility may still be utterly unnecessary.

The criterion of distinctiveness asks whether the metaphor in question makes use of distinctive features of the source field that are not easily available elsewhere. As an example of a borrowing that shows little sign of distinctiveness, consider once more Thomas Weissert's discussion of Borges's "Garden of Forking Paths" (1991, 226). Weissert builds an analogy between the multiple levels of the story and the way chaos theorists construct three "levels of reality": experimental systems, mathematical models, and graphical representation. But this scientific process is by no means unique to chaos theory; the use of levels of abstraction in modeling goes back at least to Galileo. Matheson and Kirchhoff criticize an essay on chaos theory and Tennessee Williams along similar lines, noting that it seems "utterly unnecessary" to use chaos to discuss the way Tennessee Williams can use simple characters to generate complex situations: "surely it would be equally possible for someone unacquainted with the faddish vocabulary of chaos" to recognize this (1997, 39). Yet their assertion of such a possibility does not itself render the metaphor in question worthless. When they say that "it is difficult to see why the entire conceptual baggage of chaos theory must be imported simply to note that causality is problematic, or that disorder may be regarded positively" (40), they may overstate their case. After all, no one said "must." The question of distinctiveness should be decided in the context of an overall assessment of whether the metaphor in question is helpful, not whether it is indispensable.[18]

Awareness

Drawing on the work of Leon Pomerance, Klein identifies, among the common problems associated with borrowing, "overreliance on one particular theory or perspective" and "a tendency to dismiss contradictory tests, evidence, and explanations" (1990, 88). A healthy awareness of the metaphorical status of one's borrowing helps to avoid these failings, but note that these problems afflict many kinds of research. Closed-mindedness can make for bad scholarship anywhere, not just in cross-disciplinary borrowing. Still, a healthy self-awareness can help guard against the dangers of forgetting, or disguising, the fact that one's metaphors are metaphors. Mirowski speaks strongly in favor of the need for such awareness because it makes one more likely to respond to criticisms and to be open to evaluations of consistency and adequacy: "What matters for our present purposes is not that any particular metaphor has flaws, but rather that the appropriate research community responds to those flaws in a responsible, systematic, and scientific manner, and acknowledges that metaphors have consequences for the content and conduct of inquiry" (1989, 279).

We have seen an admirable example of this self-critical awareness in the example of Michael Frayn cited earlier. In noting the disanalogies between his source and target field, Frayn exemplifies another one of Richard Boyd's principles for the use of metaphors: active consideration of the possibility that there may be no important similarities or analogies between the two fields, or different ones (Boyd 1993, 523–24). Schön points out how important criticism of metaphors is, because we already think about social problems "in terms of certain pervasive, tacit generative metaphors" and so "we ought to become critically aware of these generative metaphors" (Schön 1993, 139). As powerful as metaphors are, they can obscure or distort important features of the situation, so we need to attend to dissimilarities and disanalogies too. By bringing the metaphors to awareness, we can inquire into them critically (148–50). In other words, awareness facilitates the critical investigation of fit.[19]

Peter Mackey displays the virtue of self-critical awareness in his metaphorical borrowing from chaos theory to describe the nature of Leopold Bloom's identity in *Ulysses*. As mentioned previously, Mackey uses the image of a strange attractor to make sense of a persistent yet ever-changing personal character. Like a chaotic system, "Bloom's stream of consciousness changes and moves, yet it also exhibits an underlying form in the consistent qualities of Bloom's personality. . . . His thoughts, in fact, move like a complex system, spinning, turning, folding, almost but never quite repeating" (Mackey 1999, 119). Yet Mackey cautions us that we should not think of Bloom's personality entirely in terms of the pattern of a strange attractor in phase space, because life happens through time. "The spatial metaphor should not cloud our awareness" of the intrinsically temporal nature of the self (122). Similarly, William Demastes flags the disanalogies in his discussion of Ibsen, identifying a butterfly effect in Ibsen's *Master Builder* but acknowledging that "it roughly applies because the chaos paradigm includes elements of determinism" (1998, 78). To insist that one's metaphors be taken literally displays the corresponding lack of critical self-awareness. In addition to the examples mentioned at the end of the previous chapter, we see this also when Peter Stoicheff identifies certain texts (by such authors as Borges, Pynchon, and Nabokov) as "metafiction" and claims that they have some of the same features as chaotic physical systems. He does not say that chaos theory provides a useful vocabulary, or that there is a relation of similarity, but asserts instead, that "a metafiction text *is* a complex system" (Stoicheff 1991, 85).

An especially pointed statement of the need for self-critical awareness of our metaphors comes from Milner S. Ball, who warns us not to proclaim a metaphor drawn from our own expertise "as if it were a statement of natural, necessary fact. Our guesses and particular position are to be conscious. And self-effacing" (1985, 19). He agrees with Schön on the dangers of metaphors that have become too familiar, because when metaphors becomes solidified in place and are taken to be true they are "more dangerous because unrecognized as metaphors," and we come to lack access to alternatives (22). So the defamiliarizing function of a restructuring metaphor can reawaken our critical awareness. Ultimately, Ball endorses a pluralism about metaphors, pointing out that a metaphor is not only creative but also "restricts or eliminates or conceals. For this reason an adequate conceptual system requires alternate, even conflicting, metaphors for a single subject, and our daily living requires shifts of metaphors for fullness of thought and action" (22; see also Brown 1977, 144).

I do not pretend that there is a foolproof procedure for judging metaphors or, for that matter, judging pies. There is no checklist of absolute requirements or algorithm for assigning a numerical measure of the goodness of a metaphor. My point is just this: it is possible to criticize them, and it is often important to do so. One can make a strong case that a metaphor is not very good, or very helpful, for a given purpose. And the very possibility of responsible criticism of metaphorical borrowing implies that such borrowing is not a worthless process guided solely by whim. In this sense, by demonstrating the feasibility of criticizing metaphors I have at the same time defended their cognitive legitimacy.

7

Facts, Values, and Intervention

When researchers use chaos theory to argue for a particular economic policy, legislative reform, or ideal of literary merit, they face the vexed problem of moving from facts to values: how can we use scientific knowledge to address evaluative questions? By "evaluative questions" I mean questions about deciding what we ought to do, how we ought to arrange our society or our personal lives, or what we think makes for better or worse artistic pursuits. Those who insist on strict separation between fact and value may hold that scientific knowledge is silent on the fundamental issues at work in such questions. At the other extreme is the view that we can replace ethical, political, and aesthetic inquiry and debate by scientific inquiry—either because the evaluative questions actually get answered by science or because scientific inquiry leads us to see that the evaluative questions disappear. In this chapter I argue that scientific knowledge can contribute to answering important questions of value in that it is neither irrelevant to value inquiry nor a replacement for it. Notice that the question at hand is not the familiar issue of whether we can have "value-free science" but instead whether we should have "science-free values." Although many in science studies have dealt with the ways in which values play a cognitive role in scientific inquiry, my concern here is the way in which science can play a cognitive role in value inquiry.

In exploring the relationship between scientific knowledge and evaluative questions, I first look at those who characterize this relationship in terms of isolation, with a strict dichotomy between fact and value, and with science squarely on the side of facts. I spell out how we bridge the supposed logical gulf between "is" and "ought," and argue that facts and values are not so separate after all. In the second section I propose an alternative to both isolation and collapse in a pragmatic, limited form of naturalism. This position helps to articulate the contribution that scientific inquiry can make in addressing questions of value. My point here echoes the discussion of disciplinary pluralism in chapter 2, where I argued that different approaches to inquiry can sometimes cooperate fruitfully without requiring either strict isolation or unifying collapse. In the final section of this chapter I consider a number of ways that knowledge borrowed from chaos theory has been used to address evaluative issues about the role of governmental intervention in the economy. Here I critically examine a number of arguments about whether and how governments ought to intervene in economic and social affairs, whereas in the chapter to follow I focus more specifically on the use of chaos theory to address issues of value in literature. These investigations of specific instances in which borrowed knowledge helps bridge the gap between fact and value should recall my examination of metaphorical borrowings in chapter 6. In this case, I seek to show how the use of scientific knowledge in evaluative debate can be examined critically and should neither be rejected as impossible nor accepted as definitive.

Isolation

The supposedly inviolable barrier between "facts" and "values" often gets traced back to David Hume's *Treatise on Human Nature*, which identified a logical gap between sentences concerned with what *is* the case and sentences concerned with what *ought* to be.[1] But the dichotomy has been under fire for some time. In 1969 W. D. Hudson introduced a collection of essays on the issue with the observation that those who consider there to be a logical gulf between "is" and "ought" "are challenged today by what appears to be an ever increasing number of their professional colleagues" (13), and John Cottingham noted in 1983 that it was already a commonplace that the fact–value distinction had broken down (455). Nonetheless, we still witness mighty efforts to analyze the distinctive character of moral language, and we still hear of those attempting to bridge the gulf accused of committing some logical sin—usually the sin called the "naturalistic fallacy."

We met naturalism back in chapter 2, where it named a certain approach to science studies. What is it doing here in a discussion of evaluative argumentation, and why does it lead to a particular kind of fallacy? Well, a very strong version of naturalism would hold that a complete description of the world can be formulated using only the concepts of the natural sciences, so that all explanations are causal ones. Such a naturalistic position threatens to eliminate all talk of values.[2] In connection with this concern, naturalism usually appears in discussions about science and values in the guise of the so-called naturalistic fallacy. The terminology traces back to G. E. Moore, who inaugurated the analytic study of ethics with his account of the concept of "good." For Moore, "good" is a simple notion and cannot be defined by breaking it down into components. Indeed it is not a natural concept at all, so defining it in terms of any natural concepts is a naturalistic kind of mistake (1951, 7, 13).

Moore disposes of naturalistic definitions of good as pleasure (as well as metaphysical definitions of good as God's will, or as anything else), with his "open question" argument: when one asks "Is pleasure good?" one is obviously not wondering whether pleasure is pleasurable. Because it is always an open question whether the feature supposedly defining good is in fact good, no such definition can succeed (1951, 16). Yet the naturalistic fallacy is poorly named: Moore was targeting metaphysicians as well as naturalists, and the term does not really name a logical fallacy at all (Dodd and Stern-Gillet 1995, 735–40). Moore's entire argument for the indefinability of good rests on, among other things, a highly contentious definition of "definition." For this and other reasons, his ethical theory has fallen out of favor, but the celebrated naturalistic fallacy lives on.

Why this persistence? Well, the gulf between is and ought, the dichotomy of fact and value, and the rejection of the so-called naturalistic fallacy all serve to defend the autonomy of evaluative inquiry and especially the autonomy of ethics. Ethics often serves as the preeminent example of questions of value, and the autonomy of ethics is the doctrine that it has a distinctive subject matter and tools of inquiry, and that its results are not dictated by empirical inquiry or reducible to nonnormative terms. The concern that answers to ethical questions do not "depend on" answers from other fields may be well placed, but much depends on the sense of "depending" here. One of my goals in this chapter is to stake out some middle ground between seeing the answers to evaluative questions as simply dictated by the sciences and, alternatively, seeing them as completely independent and thus isolated from science.

Even isolationists recognize that the fact–value distinction does not imply that evidence and argument are irrelevant to morality, because practical means–ends reasoning requires empirical evidence. Just because one cannot argue *to* values does not mean that one cannot argue *from* them, by constructing practical syllogisms or by tracing the hypothetical consequences of value positions (Cottingham 1983, 459; see also Moore 1951, viii, 146). So scientific knowledge can play a non-controversial role in evaluative inquiry when it spells out possibilities and consequences for human action. And if science establishes that a particular goal or state is impossible to achieve, such a finding would enable a good argument that one is not obligated to attain such a goal or state. The philosophical slogan *"Ought* implies *can"* allows us to see that when something cannot be done, it makes no sense to say that we ought to do it. Conversely, if empirical inquiry establishes that a certain action is possible, it opens up the question of whether that action should be undertaken. We will see that such results are important in economic policy, where some policymakers contend that governmental intervention simply cannot improve an economic system, and chaos theory has been used to argue that such intervention can successfully achieve certain goals.

The autonomy of ethics (or other evaluative inquiry) can be a defense against its attempted collapse into the technical or the ideological. Isolation can serve as a strategic refuge from the onslaught of the technical, a refuge understandable but excessive.[3] Autonomy is a way of preserving value discourse from being usurped by or collapsed into other realms of inquiry. To that extent, as a pluralist I defend the autonomy of value inquiry against a scientism that seeks to simply collapse all questions into scientific ones; the results of value inquiry are not dictated by empirical inquiry. This collapse may take the form of reductionist theories that consider statements of values to be entirely replaceable by statements about consequences or feelings, for example. Such theories do not solve our evaluative problems but shirk them (Toulmin 1958, 231). Many hear a call for both disciplinary and evaluative collapse in E. O. Wilson's carefully hedged suggestion that "scientists and humanists should consider together the possibility that the time has come for ethics to be removed temporarily from the hands of the philosophers and biologized" (1980, 287). This kind of collapse may also be discerned in some borrowings from chaos theory that attempt simply to read evaluative conclusions directly from the science: for example, when Robert Scott declares that "chaos theory can best be described as a form of zen, stressing as it does the acceptance of tensions in life and physical processes" (1993, 348), or when Vincent DiLorenzo proclaims that chaos theory "finds uncertainty to be an advantage in legislative intervention" and "embraces unpredictability rather than equilibrium" (2000, 53). Of course, the science of chaos theory does not straightforwardly urge us to accept or embrace anything at all, although its results may be useful in constructing arguments that lead us to do such things. And make no mistake, collapse can also go in the other direction as well: by reducing science to nothing more than a social construction that mechanically reflects the values and ideologies of its historical context. John Lyne points out that "the error of scientism is to think that science can meet society strictly on its own terms and rewrite society's rules; the error on the other side is to treat science as if it were not a frontier and thus already fully socialized" (1990, 54).

If the collapse of evaluative discourse into science represents such a threat, why not pursue a policy of isolation? There are two reasons not to retreat behind an unbridgeable gulf between facts and values: we cross the gulf all the time, and the gulf was never as wide as it seemed. The first is a point about reasoning: there is plenty of argumentative traffic across the divide between fact and value. The second is a point about description: the purported separation between empirical inquiry and evaluative questions is not as sharp as has been claimed. So in addition to the bridges we build, there are natural features (downed trees? vines? stepping stones?) connecting the two sides of the supposedly gaping chasm between facts and values.

Looking first at the issue of argument, Stephen Toulmin has diagnosed the mistake behind the fact–value dichotomy as stemming from the belief that only strictly deductive reasoning can count as reasoning at all. In reasoning from statements about the way things are to the way we ought to act, there is indeed a change in logical type that makes the tautological validity of analytic inference impossible. But few if any real-life arguments are as elegantly tidy as analytic arguments, and they can be conclusive nonetheless (Toulmin 1958, 125). When philosophers claim that one can never reason from facts to decisions, they are relying on an excessively narrow notion of what is to count as "reason," for "no collection of statements about our present situation, the consequences of our actions, or the moral scruples of our contemporaries and fellow-citizens can entail a conclusion about our obligations" (222). If one holds to the analytic ideal and insists on an analytic solution, one is simply not facing the problem: there seems to be a "logical gulf," but it is a gulf that each of us crosses regularly (167). Instead of obsessing about our failure to cross this gulf using only tools guaranteed to fail, we should instead ask about how we do, in fact, reason from "is" to "ought" and examine the ways in which such reasoning works better or worse. Here, as in previous chapters on rhetoric and metaphor, I offer a brief theoretical defense of the possibility of this type of critical attention and then proceed to demonstrate that possibility by actually critiquing examples of borrowed knowledge that construct bridges from facts to values.

The second reason we can make our way from facts to values is that we never faced a gaping chasm in the first place. Descriptive enterprises, which may seem to trade only in facts devoid of evaluative character, actually depend on values in order to operate at all; there can be no "value-free science." Trade-offs in reconciling theory and observation rely on values like coherence and simplicity; they are values that genuinely guide actions and are not mere expressions of subjective emotion. Without these cognitive values we would never have scientific facts (Putnam 1990, 138–39). We need not here join the debate about whether social, external, or "contextual" values enter into science, although the point is made forcefully by Helen Longino (1990), among others.

Furthermore, descriptions can themselves involve evaluation as Hilary Putnam has argued. Some terms simply ignore the fact–value dichotomy; these "thick" concepts such as "crime" or "cruel" require us to enter into an evaluative community in order even to make sense of their ascription (Putnam 2002, 35–38). When it comes to these "thick" evaluative concepts, there is no separating out the descriptive component from a prescriptive component; fact and value are entangled in the very notions. In making these observations, Putnam draws on the work of Iris Murdoch, who holds that the dichotomy comes from a mistaken moral psychology—where uncaring reason finds neutral facts, and then the will chooses values either arbitrarily or based on instinct (Putnam 1990, 150). "Our real-life world, Murdoch is telling us, does not factor neatly into 'facts' and 'values'; we live in a messy human world in which seeing reality with all its nuances . . . and making appropriate 'value judgements' are simply not separable abilities" (166). Real arguments about real policy questions exemplify "the entanglement of the ethical and the factual"; actual ethical arguments never involve a tidy agreement on facts and disagreement on values (167).

Yet a gulf, however bridgeable, still presents a convenient geographical boundary. Descriptive and evaluative enterprises are not the same, and we should resist attempts to collapse them. As Putnam puts it, there is no *dichotomy* between fact and value, but there is still a meaningful *difference* between them (2002, 11). Because we cannot simply dismiss those who use science to argue for evaluative claims, we must do the work of examining their arguments to see how sturdy the proposed bridge will be and whether it will bear the traffic that is asked of it. In saying this, I reject the notion that policy decisions, moral judgments, and evaluations of aesthetic merit are merely expressions of personal preference or cultural prejudice. In at least some circumstances it must be possible to reach intersubjectively valid evaluative conclusions based on consensually developed standards of judgment. Sometimes, one can indeed account for taste.

Pragmatist Naturalism and Borrowing across the Valley

When we consider the interactions between facts and values and between scientific inquiry and evaluative questions, the pragmatism of John Dewey provides an alternative to both isolation and collapse. Dewey was in many ways a naturalist who advocated an idea that now sounds outrageous: that ethics should become scientific, and value inquiry should use scientific methods. In this section I briefly sketch Dewey's notion of values and value inquiry. A consideration of Dewey's naturalism and its treatment of the relationship between facts and values provides the framework for the detailed examinations of borrowing to follow.[4]

In his essay "The Logic of Judgments of Practice," Dewey outlines his view that values are not elements of a pre-given reality to be discovered. Instead, we construct them as a result of interaction and judgment. We can act toward something *as if* it is of value, by prizing and delighting in it—an immediate, subjective response. But judging that it actually is of value is different, because judgments of value are practical judgments about actions. Acting as if something is food means to eat it, but judging it to *be* food means finding that it actually nourishes by looking at the consequences of eating it (Dewey [1915] 1979, 26–29). This approach to the analysis of consequences differs greatly from certain utilitarian accounts that simply take our pleasures and pains at face value. Indeed, much of contemporary economics claims to abjure normative questions yet takes the satisfaction of desire as an unqualified good without inquiring about where those desires came from or whether they genuinely conduce to well-being (Dupré 2001, 3; on this point see also Sen 1987).

Not all things that are valued are in fact valuable, and Dewey held that systematic inquiry was required to discern and articulate our values. For Dewey, many problems persist because of the view that moral issues can be settled by consulting private conscience rather than by studying and applying knowledge. In addressing the tasks of improving the social order and making possible the cooperation needed for human flourishing, ethics must make use of the methods and results of the social sciences (Dewey and Tufts [1908] 1978, 8). Dewey insists that his point of view "expressly disclaims any effort to reduce the statements of matters of conduct to forms comparable with those of physical science. But it expressly proclaims an identity of logical procedure in the two cases" (5). Note that for Dewey, "logical procedure" means something like "epistemic procedure," so his suggestion that ethical claims should be treated scientifically does not imply that we should "turn moral philosophy into an applied science" (Ruse and Wilson 1986, 173) but just that we should use systematic and experimental procedures for justifying judgments.

Because value inquiry represents one form of inquiry in general, it can yield knowledge every bit as much as scientific inquiry can. Questions about the design of social institutions have answers just as objective as do those about the design of bridges: one must look at the situations given, the purposes to be achieved, and the relative merits of different designs (Welchman 1995, 1–8). Dewey's pragmatism rejects the notion that we achieve objectivity by a faithful picturing of preexisting reality and instead requires us to achieve objectivity by means of—among other things—open criticism and honest evaluation of practical endeavor. Indeed, Hilary Putnam has diagnosed part of the appeal of the fact–value dichotomy to be that it lets us off the hook from having to do the hard work of examining our convictions; we can simply declare them to be subjective preferences immune from criticism (2002, 44). And it is precisely because value inquiry yields knowledge that I include the critical examination of evaluative arguments as part of my project of examining cross-disciplinary borrowing from the natural sciences.

We see, then, that pragmatism involves a certain kind of naturalism in value inquiry. But this naturalism resembles naturalized epistemology (which sees philosophy as continuous with empirical inquiry) rather than naturalistic ethical theories (which seek to collapse evaluative discourse into the descriptive). Where is the "nature" in this naturalism? Well, Dewey believed that "solutions to the problems of promoting social cooperation were to be found in the study of human nature" (Welchman 1995, 1). That is, value inquiry starts from facts about us as human beings.[5] But although it *starts* from facts about us, it does not *stop* there. Even Michael Ruse and E. O. Wilson seem to recognize this important distinction at one point in their article "Moral Philosophy as Applied Science," which serves in many ways as a call for collapse. They acknowledge that scientific explanations do not automatically lead to moral codes, even though they underlie moral reasoning (1986, 190). Instead of attempting to simply read ethical conclusions off scientific facts about human beings, pragmatist value inquiry relies on the interaction of a number of strands: empirical inquiry into human biology, psychology, and sociology as well as critical reflection about concrete endeavors (Welchman 1995, 145, 190). The question then is not whether it is possible to use scientific facts to argue for evaluative positions. Instead, we face the task of critically examining the ways that borrowed knowledge is used to construct argumentative bridges from the descriptive science of chaos theory to evaluative claims about economics, law, and literature. These bridges may rely on metaphor, on "thick" descriptive terms, or on substantive notions of human nature. I turn now to a brief consideration of these materials for spanning that gulf.

Metaphor

As we have seen in the previous chapter, structuring metaphors can help to organize inquiry by providing both theoretical vocabulary and hypotheses to pursue. But Donald Schön points out that metaphors can also provide a structuring narrative that frames a problem situation and leads to normative prescriptions. Such metaphors can pick out a determinate problem from a complicated social situation, by selecting salient features and providing a coherent description. This framing helps make the leap from "is" to "ought," making it seem obvious (Schön 1993, 146–47). The metaphor of urban blight as disease sets up a whole way of thinking: we need to cure it, prevent a relapse, find the root cause or etiology. Once one casts something as a disease, it becomes evident that one needs to cure it.

Consider the thinking of those framers of the Constitution who conceived of it as a Newtonian machine with counterbalancing forces (Tribe 1989, 7). Here we see a structuring metaphor that bestows upon us the problem of constructing a mechanism that continues to run smoothly. Such a metaphor carries with it a normative presumption that one wants the machine to keep running smoothly, so the forces must be balanced harmoniously. Especially with functional concepts, once the metaphor is constructed in terms of health and disease, or smooth operation versus malfunction, normative force is built right in. Lakoff and Johnson use the example of the competing metaphors "love-is-madness" versus "love-is-collaborative-artwork" to point out that structuring metaphors can "sanction actions, justify inferences, and help us set goals" (1980, 142).

In the case of a metaphorical borrowing used to support an evaluative claim, the process of critique has two stages. First, the metaphor can be examined using the criteria outlined in the previous chapter to see if it is strong, promising, fruitful, and so forth. Such a critique will focus on the question of whether the scientific concepts, methods, or results of the source field prove genuinely useful for describing and understanding the specified area of the target field. Second, we can investigate the distinctive issues that arise in justifying an evaluative claim. These issues might include the following: What is the community of inquiry for which this claim is held to be valid? How have the evaluative criteria been developed and agreed upon? What carries us from descriptive adequacy to evaluative judgment in this particular case? Questions such as these aim to reveal the argumentative underpinnings for the proposed bridge from description to evaluation. Such bridges are built every day, and we can examine them critically to see which ones are strong enough for us to travel upon.

Thick Descriptions and Value-Laden Terms

Perelman and Olbrechts-Tyteca point out that some terms, such as "light, height, depth, full, empty, and hollow," have been used so often in analogies with the moral realm "that it is no longer possible to detach from them the value derived from this role, as a consequence of interaction with certain terms of the theme" (1969, 382). Such metaphorically loaded terms, especially notions of what is "stable" or "flexible," show up frequently in discussions of chaos theory (to say nothing of the term "chaos" itself). Indeed, some terms are so metaphorically loaded with values that they approach the status of such "thick" concepts as "cruel" or "crime," which cannot even be used correctly without participating in a community enterprise that is at once descriptive and evaluative. Knowledge borrowed from the natural sciences often traverses the distance between description and prescription by means of just such terms. Consider the term "healthy" and its contrary, "pathological." Historian of science Timothy Alborn notes that the notion of the healthy body was central to nineteenth-century social science, and we easily notice that this has hardly changed (1994, 173). Biological scientists have observed that a number of well-functioning physiological systems display chaotic behaviors (Goldberger 1991). When these descriptions of chaos-as-health migrate metaphorically to economic or legal systems, they allow the drawing of prescriptive conclusions, as, for example, when Robert Scott argues that apparently random fluctuations in the legal system should be accepted as healthy (1993, 351).[6] In economics, William Baumol and

Jess Benhabib begin their review article on the role of chaos with the invitation to consider a narrative marked by thick descriptions:

> Imagine a bargaining model (say, involving diplomats negotiating tariff levels or a disarmament treaty) in which each party has been instructed by higher headquarters to respond to each new offer by her opposite number with a counteroffer that is to be calculated from a simple reaction function provided in advance. Both negotiating parties are prohibited from revealing their own reaction functions to the other. If the perfectly deterministic sequence of offers and counteroffers that *must* emerge from these simple rules were to begin to oscillate wildly and apparently at random, the negotiations could easily break down as each party, not understanding the source of the problem, came to suspect the other side of duplicity and sabotage. Yet all that may be involved, as we will see, is the phenomenon referred to as *chaos*, a case that is emphatically not pathological. (1989, 77, emphasis in original)

Economics is marked by its use of the evaluatively charged concepts of stability and efficiency. Although these notions may not fully qualify as "thick" descriptive terms, they often play as crucial a role as terms like "pathological" in mediating between facts and values. As Putnam has pointed out, even descriptive terms can have an ineliminable ethical flavor to them (2002, 63). Recall the discussion in chapter 4 about the ways in which chaotic behavior was neglected for quite a while. We noted Randall Bausor's explanation that one cannot study chaotic behavior without paying attention to instabilities: "By inclination and training, however, economists abhor instability.... To most economists competitive processes that rule the economy are inherently dynamically stable.... Their most cherished attitudes toward markets and their most central presumptions about how the economy should be governed are all profoundly challenged by analyses conditioned on systematic instability" (1994, 122). Thus, randomness was traditionally modeled with exogenous shocks. Indeed, when Paul Samuelson's classic economic textbook discusses the cobweb model of a time-delayed situation in which there is dynamical change in prices, he says that this model "throws some light on the business-cycle phenomena . . . where a free enterprise system would tend to fluctuate if not moderated by public policy and equilibrating market mechanisms. Firms, like farmers, can swing from one extreme to the other, thus causing instability"

(1980, 381). Notice that instability is portrayed here as a problem to be solved and that this very cobweb model would soon become one of the places where chaotic behavior was first studied in economic systems (Jensen and Urban 1984). But why is instability so bad? And how can we reach any easy conclusions about whether and how to pursue stability when the scientific literature offers a number of different conceptions of stability to chose from—especially the difference between the asymptotic stability that resists small perturbations to a particular trajectory and the structural stability that resists the effects of small parameter changes on the system's long-term type of behavior? Chaotic systems can show a great deal of structural stability, after all, even though all their trajectories are highly unstable to perturbations.[7]

In contemporary economics, evaluative concerns are often packed into the notion of Pareto optimality, which roughly describes a situation in which no one's position can be improved without making someone else's position worse. Far from being a "thick" descriptive term, Pareto optimality seems especially thin because it excludes consideration of substantive values such as justice while facilitating the application of technical analysis.[8] The field of economics is usually strictly divided into the positive and the normative, with the positive simply describing economic facts and laws for policymakers to use in decisions. Normative or "welfare" economics has shrunk to the point that it mainly focuses on Pareto optimality, because that notion describes precisely a situation in which there can be no disagreement about whether it is best. How did it come to this? Amartya Sen describes the situation as follows: when values are considered merely subjective, substantive rational inquiry into values becomes impossible and Pareto optimality becomes the only criterion left for evaluative assessment of economic functioning (1987, 31; see also Putnam 2002, 54).

In pursuit of this form of optimality, most positive economics simply assumes that efficiency is a worthy goal rather than engaging in argumentation about what economic goals should be (Dupré 2001, 146–47). The term "efficient" carries so much evaluative weight in economics that it sometimes occludes all other evaluative discussion. Samuelson links the notions of efficiency and health when he warns against such policies as rent control and the minimum wage by stating that "the pathology of interference with supply and demand helps to bring out the remarkable efficiencies produced by perfect competition when it is able to operate" (1980, 371). Yet Samuelson must count as a moderate because he admits that loss of efficiency for equity is sometimes acceptable. As we will see, the evaluative weight folded into the concept of economic efficiency makes it—if not exactly "thick"—ample enough to help construct bridges from descriptive results to policy prescriptions.

Human Nature

As mentioned earlier, Dewey's naturalism welcomed inquiry into human nature as a source of important information for constructing our value systems. Indeed, John Dupré states that it is "a commonplace that no normative political philosophy can get off the ground without making some assumptions about what humans are like" (2001, 86). Much recent investigation of human nature has taken the form of an evolutionary psychology that sees human nature as the result of our long biological heritage. Although some may be tempted to condemn this enterprise a priori, a pluralist and naturalist approach should consider that our decisions about how to arrange our personal and societal affairs might well benefit from a better understanding of the moral sentiments that humans actually have. Descriptive investigation of these sentiments should be able to take place in the context of an evolutionary framework and could yield relevant information, even if most examples of actually existing evolutionary psychology are flawed or even worthless.[9] We should not simply screen off our biological features when seeking to understand ourselves and make a better world, but we should remain vigilant and even suspicious of attempts to spell out constraints that seemingly come from nature, because there is such a long history of humans fooling ourselves into believing that contingent social formations are supported by such natural constraints.

The work of rhetorical analyst John Lyne provides a valuable resource in the critical examination of bridges built with appeals to human nature. Lyne seeks a path between some analytic philosophy, which closes off connections between "is" and "ought" logically, and some critical theory of ideology, which makes the connections seem inevitable and mechanical (1990, 36). He identifies a "bio-rhetoric" as calling attention to "the potential for borrowing rhetorical resources from one domain and using them in another. Such a strategy makes biological considerations discursively available to discussion of social, political, or moral issues, or vice versa—the flow can be in either direction," and in this cross-fertilization, "heuristic resources from one discursive domain enter and animate another" (38). The suggestion that we put our thinking about morality and politics on a biological foundation that respects our evolutionary limitations runs into "the paradox inherent in suggesting, in effect, that we should strive to make things conform more to their nature" (43). But "the claims of the theory of human nature can be aligned to a variety of social conclusions, depending on the path of rhetorical invention." Do we adjust to our biological nature or fight against it? The strategy of complementarity embraces our nature, whereas the strategy of compensation challenges it. Both are ways to "quietly naturalize moral judgment" (44–45).[10] So considerations of human nature based on our biological heritage can play a role in evaluative inquiry, along with metaphors and "thick" descriptive terms that carry prescriptive weight. None of these elements should be ruled out of bounds as a possible component of argumentative bridges from facts to values, but each of them must be subject to careful critical scrutiny.

The Value of Intervention

In the remainder of this chapter, I examine one particular arena in which bridges have been built from facts to values: questions about the role of governmental intervention in economic affairs. First, I explore arguments that use chaos theory to make a case for policies seeking to counteract wide economic fluctuations, as well as arguments that chaos theory proves the futility of all governmental intervention. Next, I consider challenges to the efficient markets hypothesis that borrow from chaos theory to advocate new directions in governmental regulations. And although the issue of intervention arises most prominently in economics, I conclude with a discussion of some of the ways chaos theory has been used to support evaluative stances in legal theory and even in literature. My goal is not to resolve all of these contentious policy issues but to demonstrate that it is possible both to borrow knowledge from the natural sciences in order to make evaluative arguments and to critically evaluate such arguments using the tools developed in this and previous chapters.

Chaos Theory and Countercyclical Policies

Speculations on the possibility of tempering wild economic fluctuations and rendering the economy more stable arose with some of the first inquiries into the possibility of chaotic behavior in economic systems. For this reason, questions about the possibility and advisability of governmental intervention provide an especially apt case for examining the ways bridges are built from results in the sciences to policy prescriptions.

Arguing for Governmental Intervention. One of the first proponents of using chaos theory to argue for countercyclical intervention was Jean-Michel Grandmont, whose 1985 work demonstrated that a competitive, laissez-faire economy can undergo chaotic fluctuations without external shocks. He studied a model of the economy with two overlapping generations and found that cycles emerge when older agents are much more risk-averse than the young and have a higher marginal propensity to "consume" leisure. Grandmont showed that countercyclical monetary policy is highly effective in this model—the government can intervene by transferring money and stabilizing expectations about the rate of interest, leading the economy to a steady state. Yet he recognized that governmental policies can generate fluctuations as well as stabilize them (Grandmont 1985).[11] As described in chapter 3, many researchers have since done the theoretical work of expanding on the initial results of Grandmont, Benhabib and Day, and others. This recent work has examined how constraints such as timescales or degree of competition can be relaxed while still allowing for chaotic dynamics in the models. Chaotic behavior has appeared within models with ever more realistic assumptions and parameter settings, and policy regimes have been outlined that would damp down the chaos in such models. For example, Michael Woodford uses his model of financial intermediation to show that the government cannot stabilize the economy using only the distribution of information and would still need traditional interventions such as fiscal policy of taxation and monetary transfers (1989, 331). Christiano and Harrison use a nonlinear model to show that an "automatic" stabilization tax policy can yield significant gains in welfare, in part by helping to dissuade people from their beliefs that market participation is beneficial during a speculative bubble. Yet they note that "most mainstream equilibrium models suggest that, at best, the gains from macroeconomic stabilization are small" (1999, 4).

Richard Goodwin, a pioneer in the use of nonlinear models to study economic oscillations, has used chaotic dynamics in a model of economic growth. He writes that "compensatory fiscal policy may appropriately be regarded as a simple type of dynamic control implemented at discrete time intervals. It aims, or should aim, to diminish the amplitude of fluctuations in economic activity" (1993a, 52). Such control is difficult, however, because of the time lag in collecting and analyzing data and the continual need to tune variables and parameters in order to approach the desired state gradually. Yet flexible dynamic control can work: "that such a fiscal policy can be, in principle, astonishingly successful is thus demonstrated. And that such result is highly desirable is also clear" (54). Unfortunately, Goodwin notes that the success of the intervention policy within his model is achieved with 10 percent unemployment and large surpluses in government revenue, which would be undesirable and unpopular. So he admits that his example is simplified and impractical but insists that "what is not impractical is the basic logic of compensatory fiscal policy" (56).

Such logic relies on the notion that large and unpredictable economic fluctuations should be minimized if possible, building a bridge from nonlinear dynamical models to policy prescriptions on the basis of evaluatively charged notions such as optimality and stability. Woodford exemplifies this argumentative pattern when he states that "the fluctuating equilibria shown to occur for some parameter values [of his model] are plainly not desirable, because any Pareto optimal allocation of resources must . . . converge asymptotically to a constant capital stock and a constant level of output" (1989, 331). Notice that Pareto optimality serves as the standard for evaluation here—as it does throughout much of economics—and that a Pareto-optimal situation (in which no one's lot can be improved without harming someone else) requires stasis; any change must imply that someone is being hurt. If fluctuations arise from the internal dynamics of the system (as discussed in chap. 4), then it becomes vital to extinguish them if possible. But as we also saw in chapter 4, economists still actively debate the question of whether we have empirical evidence of chaotic behavior in macroeconomic data. Brock and Malliaris note that those holding the more traditional view that fluctuations arise from external jolts typically argue that governmental interventions are futile at best. They write: "we do not think that anyone in macroeconomics needs to be convinced of the drastic difference in policy implications" of the view that fluctuations might arise endogenously— proponents of such a view typically suggest strong policies to stabilize such fluctuations (Brock and Malliaris 1989, 305).

Questioning the Implications of Chaos Theory. Of course, not everyone agrees with those economists who use chaos theory to argue for stabilization policies. Peter Schwefel asks what deterministic chaos in macroeconomic data would even mean for policy: "Does it mean that economic policy can't do anything because we don't know the initial conditions? Or would that mean we should set up some rules and framework for economic policy within which the economies operate?" (Day and Chen 1993, 319).[12] James Bullard and Alison Butler, two economists at the Federal Reserve Bank of St. Louis, have explored in depth the policy implications of chaos theory. They argue that the nonlinear models used to justify a role for interventionist policy typically include some deviation from perfect markets such as incompleteness, expectations that are not fully rational, or externalities.[13] Discussing a model of capital accumulation that exhibits chaotic behavior, they note that this behavior is Pareto optimal, so "there is no role for stabilisation policy, as government intervention cannot lead to a Pareto-superior outcome" (Bullard and Butler 1993, 856). Notice that Bullard and Butler assume that a failure of Pareto optimality is the only way to justify intervention. They point out that Grandmont's 1985 model of endogenous chaos justified countercyclical intervention because the chaotic price behavior made perfect foresight impossible. But if there is not perfect foresight, there is departure from a perfect market situation, so that in itself justifies intervention. Models that start with a situation of imperfect markets can offer no new argument against laissez-faire or for intervention because they do not provide an example of Pareto-inferior chaos in a perfect market (Bullard and Butler 1993, 858–59). Yet they still see an important role for nonlinear dynamics in economic policy: "Even if nonlinear dynamic models offer no new justification for policy, it still remains that linear models may be misleading in a fundamental way, skewing our understanding of the economy and corrupting the associated policy advice" (859). For example, in a hyperinflation model the assumption of linearity clouds the conclusions because there may be more than one rate of inflation at which stabilization can occur (862). And if stabilization is justified, the nonlinear models include parameters that make a big difference to the behavior. "When some of the parameters of the model can be set by the policy authorities, the authorities have considerable control over the dynamic outcome" (850). So chaos can indicate the efficacy of intervention, even if some theoretical models provide no new justification for it.

An even greater challenge to those who would use chaos theory to justify policies of intervention is presented by J. Barkley Rosser Jr. in his book *From Catastrophe to Chaos: A General Theory of Economic Discontinuities* (1991). One problem with policy designed to damp down economic fluctuations stems from the steep requirements for knowledge needed to intervene effectively. Rosser notes that Grandmont's optimism about governmental intervention requires "not only that the government knows what it is doing, but that the economic agents *believe* that the government knows what it is doing, a tall order indeed" (1991, 115). In light of the fact that chaotic fluctuations can undermine the learning necessary for successful intervention, Rosser considers the benevolence and omniscience required for "New Keynesian optimism" to be "naïve" (115, 121). Thus, governmental intervention is risky: "Especially in light of the problem of the sensitive dependence on initial conditions, the probability that any attempted government policy will fail to achieve its target by a wide margin is certainly significant under conditions of incipient chaos" (321). It gets worse: Rosser argues that intervention in chaotic systems may be not only difficult but actually harmful. He notes that one of the earliest models of economic chaos, from the 1982 work of Benhabib and Day, shows that monetarist interventions can actually lead to chaos, subverting the usual prescriptions for policy (Rosser 1991, 112).[14] He goes on to cite Brock as saying "policy noise may be worse than market noise" and concludes one of his chapters by saying, "not only does chaos breed confusion, but chaos breeds more chaos," thus washing his hands of policy prescriptions (122). But we must distinguish between intervention directed at the level of the exact details of the system and intervention aimed at adjusting large-scale parameters. Failure to make this distinction constitutes a fatal flaw in the argument of Don Lavoie, who borrows from chaos theory in order to make a case against any governmental intervention whatsoever.

Arguing against Governmental Intervention. According to the Nobel Prize–winning economist Friedrich Hayek, central planners face insuperable problems of calculation because deciding how much of each good should be produced would require the solution of hundreds of thousands of simultaneous equations each moment, and similarly astonishing calculative feats would be required for setting prices (1948, 156; see also 182, 187). This "calculation argument" of Hayek, Ludwig von Mises, and others may seem at first remarkably similar to certain findings from chaos theory. Some have argued that chaos theory proves the optimality of laissez-faire capitalism, but this contention ignores alternative and equally plausible conclusions. Hayek argued against "the belief that deliberate regulation of all social affairs must necessarily be more successful than the apparently haphazard interplay of independent individuals" (as quoted in Lavoie 1989, 622). But note that this quotation makes a fairly weak claim, denying that total planning is necessarily superior. Even granting that chaos theory supports this claim does nothing to establish the much stronger proposition that the interplay of free individuals is more successful than any kind of planning whatsoever (see also Hodgson 1994, 438).[15] Although some who discuss chaos seem to see it as indicative of problems with laissez-faire systems, economist Don Lavoie takes a position akin to Hayek's arguments against planning and looks at chaos as the key to self-organization. Lavoie invokes chaos theory as an antidote for the mechanistic physics that inspired some strands of orthodox Marxism and attempts to show that centralized planning of every detail of the economy is "utterly futile" (1989, 615). In this he may well be correct, but he goes on to make the stronger claim mentioned earlier. He (together with some other radical free-market theorists) uncritically assumes that a free market will "self-organize" to avoid inflation and unemployment.

Talk of self-organization refers to the Prigogine school of nonlinear dynamics and the related work on complexity theory at the Santa Fe Institute. As Geoffrey Hodgson has pointed out, Hayek attempted to link his work with scientific ideas such as self-organization and spontaneous order (1994, 432). Yet Hayek and Lavoie ignore the equally valid possibility of the spontaneous *disorder* represented by economic chaos, which would seem to provide some grounds for interventionist policies. There is no guarantee that systems always evolve toward an ordered state (Hodgson 1994, 433). The closest to laissez-faire we have seen, in the United States and Great Britain around the turn of the twentieth century, had great fluctuations, inequality, and insecurity (Rosser 1991, 322). Yet Lavoie suggests that the disruptions and suffering caused by capitalism are due to "attempts to centrally control the money supply" (1989, 623). He writes that "the severe problems with inflation and unemployment that plague most Western capitalist economies are not due to the fact that they have relied too much on spontaneous market forces, but rather to the fact that they too have tried to control rather than cultivate the economic order" (632). And he suggests that nonlinear dynamics provides an argument not only against centralized planning but against even the economic intervention associated with our familiar "mixed economies," that is, "the ideas about 'fine-tuning' the economy by management of the money supply and fiscal policy, which have come to be known as Keynesian" (628). According to Lavoie, "the fine-tuners know as little about what they are doing as do the central planners" (628). But this conclusion is in no way established by the science of chaos.

In fact, chaos theory casts doubt on such free-market enthusiasm. Even a situation of distributed economic agents with perfect local knowledge can lead to chaotic price behavior. And the inability to make exact predictions of the price for each and every commodity does not entail the inability to judge what realms of "parameter space" are probably more beneficial. Consider the following analogy between the economy and the weather: The earth's atmosphere is a highly complex, spatially extended system that is coupled to a number of other systems. Nonetheless, we can construct an argument that the weather is probably not predictable in detail beyond a short span of a few weeks. On the basis of some simple models such as the Lorenz system and some plausible assumptions relating them to the behavior of the atmosphere as a fluid-dynamical system, we have good reason to believe that the atmosphere's dynamics has chaos embedded within it. We could construct a similar argument to the effect that a nation's economy has chaos embedded within it. Such an argument would turn on the presence of chaotic behavior in some simple economic models, together with an argument that it is plausible to believe that these types of mechanisms are present in the actual economy at least somewhere. The links between these chaotic models and the actual economy may well require more questionable assumptions than those linking the Lorenz system and the atmosphere, but imagine that such an argument can be constructed.

The argument against large-scale economic planning presents itself now as the analogue of the following argument: Because the earth's weather is unpredictable in detail, we should do nothing about the greenhouse effect. We have no way of knowing with certainty whether the global warming observed over the past hundred years is a random fluctuation, and we have no way of being sure about the effects of lowering carbon dioxide emissions. Any attempt to change the atmosphere for the better is destined to fail.

Clearly, this quietistic argument is fallacious. The inability to predict a system in detail does not imply the hopelessness of intervention. In fact, the tools of chaos theory allow us to study systems in various realms of parameter space, to see how the overall behavior responds to external changes in crucial parameter settings. Lavoie seems to admit as much when he says that we can "exert influence over the workings" of a process even if we cannot control its "detailed working" (1989, 619, 621). But according to Lavoie, this "cultivation" of beneficial economic order must be restricted to the creation and maintenance of unfettered markets, because any more ambitious attempt to control an economic system would "be more likely to interfere with its own logic and obstruct its self-ordering than to intelligently 'guide' it in any sense" (621). Despite this assertion, the fact that we could never control a chaotic economy in precise detail says absolutely nothing about the possibility or advisability of coarse-grained macroeconomic adjustments such as those pursued by monetary policymakers, which I discuss later in greater detail.

As I have argued elsewhere (Kellert 1993, chap. 4), chaotic behavior is unpredictable in quantitative detail but often highly predictable on the qualitative level. Therefore, even assuming that the economy is chaotic does not render successful interventions impossible. Such interventions would have to disavow the hubristic assumption that exact predictability is possible and instead aim to discover general patterns of desirable behavior and "cultivate" them (see Cartwright 1991, 45, 53). Chaos theory does not break with the Western scientific tradition of the objectification and control of nature (Kellert 1993, 115). Developments in "chaotic control" bear out this idea; researchers at the United States Naval Surface Warfare Center have achieved the control of a chaotic experimental system (Ditto, Rauseo, and Spano 1990). By using the techniques of nonlinear dynamics, these researchers were able to locate the unstable periodic orbits that are known to litter the phase space of a system with chaotic behavior. By persistently applying small perturbations to an experimentally accessible parameter of the system, they are able to guide it to an orbit of a desired period and keep it there. In this way the very techniques of nonlinear dynamics allow some chaotic behavior to be delicately tuned toward a specific periodic regime. Sensitive dependence can surely subvert attempts to master every last economic variable, but the techniques of "chaotic control" can be used in some models, even without complete knowledge of the system (Kaas 1998, 314). We cannot conclude from the possibility of economic chaos that all intervention is impossible.

What, then, are we to make of the ambivalent messages that chaos theory seems to bear for governmental intervention? The situation may prove not so different from some earlier metaphorical borrowings, such as the image of the circulation of the blood. Timothy Alborn argues that this metaphor was especially useful in debates in early Victorian economics precisely because of its ambivalence: one could conjure the image of either steam circulating in a mechanical system or blood circulating in an organism. The first lent itself to a discourse of centralized administration, the second to laissez-faire (Alborn 1994, 176). In general, images from the natural sciences have no simple political valence even when they become idolized as a result of scientism. As Mirowski points out, Hayek denounced scientism as leading to socialism and serfdom, but "physics metaphors have been at least as instrumental in reifying the image of a natural self-regulating market as they have been in encouraging engineers to believe in their own capacities to successfully plan economic activity" (1989, 356).

Rosser provides a similarly evenhanded conclusion: chaos theory shows us that central planning can be too rigid and laissez-faire can be too unstable. Yet moderate interventions carry no guarantees that they won't make it worse. "In the face of potential catastrophe and chaos, every economic system faces difficult challenges that cannot be resolved by mere rhetoric, sloganeering, or blind ideological faith" (Rosser 1991,

323). Lest this sound like a counsel of despair, we should note that these challenges can still be met. We can, after all, rule some things out, like exact total planning—and we can safely say that assertions about the impossibility of chaos are wrong. Markets do not always reach stable equilibrium, and we can get some good ideas about how tuning parameters might work. The fact that borrowings from the natural sciences do not provide all the answers to our policy questions does not mean that they have nothing relevant to say.

Efficient Markets

What Is the Efficient Markets Hypothesis? A central tenet in the defense of capitalist economies against central planning is the notion that the competitive price-setting process represents the best, if not the only, way to summarize and encode the information needed to assign a value to each good. As Lavoie puts it, "participants to the market process contribute to the discovery and conveyance of knowledge by imparting their local information to prices and in turn receiving useful information from others that is digested in prices" (1989, 629). Philip Mirowski states that neoclassical rational expectations theory, which holds that current prices reflect all essential information, has largely supplanted Keynesian macroeconomics and argues against the need for any major governmental intervention (1994b, 472). After all, when all relevant information is conveyed by prices, markets can distribute all goods and services with perfect efficiency, and governmental interference can only lessen the people's welfare.[16] The economist Lawrence Cunningham has addressed the implications of chaos theory for this so-called efficient markets hypothesis at some length in a law review article that draws some striking conclusions for governmental policies and regulations. Cunningham documents the way the efficient markets hypothesis (EMH) dominates debates in corporate and securities law and policy in part because of its presumably scientific character, and he notes that the Securities and Exchange Commission has relied extensively on the EMH in developing rules about such issues as securities registration and shareholder disclosure (1994, 548–50).

As laid out by Cunningham, the recent history of the EMH begins with empirical studies from the 1960s that found no patterns in the price history of securities (items such as stocks and bonds). The lack of significant correlations seemed to indicate that past price data were of no help in predicting future price behavior.[17] As a result of the failure to find predictable patterns, economists adopted a "random-walk" model of securities prices wherein tomorrow's price did not depend at all on previous prices. But how to make sense of a market that moves in such a completely random fashion? The EMH was propounded to explain this random model by hypothesizing that in a theoretically perfect market (where everyone is fully informed and completely rational), prices adjust instantaneously to reflect new information, the sources of which are assumed to be random (Cunningham 1994, 551–59).[18]

Cunningham spells out three forms of the efficient markets hypothesis: a weak form, a semistrong form, and a strong form.[19] The weak form maintains that "current security prices fully reflect all information consisting of past security prices," the semistrong form that "current security prices fully reflect all information that is currently publicly available," and the strong form that "current security prices fully reflect all currently existing information, whether publicly available or not" (Cunningham 1994, 560). Notice that the strong form boldly claims that one cannot systematically make money even with insider information, a claim so soundly disproven by the insider trading scandals of the 1980s that we need not deal with it further (562).

Let us trace out the chain of implications for the weak form of the EMH. If the current price of a security conveys in digested form all the information contained in its price history, then any change in price must come from new information. If the source of this new information is assumed to be random influences from elsewhere in the socioeconomic system, then the price will follow a random walk. And if the price behavior is random in this way, then it will be impossible to predict the future price behavior of the security on the basis of its price history. Thus, the weak form of the EMH explains the systematic failure of all so-called chartist analyses, or technical studies of price data, as an aid in prediction (Cunningham 1994, 561). As a rough framework for investors, the EMH simply offers the excellent advice that one cannot beat the market in the long run. But as a scientific hypothesis it yields a chain of implications that we can trace as follows: If securities prices were to show dependencies on their history, then they would not follow a strictly random walk. And if the price does not follow a strictly random walk, then changes in the price would not be the exclusive result of new information being gained. This chain of implications is the key to one of the main challenges to the EMH described in the following section.[20]

Problems with the EMH. As we have seen, the EMH was posited as an explanation for securities price histories' failure to display patterns or correlations. But those testing for the randomness of price data were looking for linear relationships by the use of tools such as autocorrelation that cannot find nonlinear dependencies. Indeed, most empirical tests of the EMH have been based on linear models (Barnett and Serletis 2000, 704; Cunningham 1994, 558). As discussed in chapter 4, the search for chaotic behavior in economic data remains an area of active research. Yet there is "clear evidence of nonlinear dependence" in financial data (Barnett and Serletis 2000, 715; see also Cunningham 1994, 573), including minute-by-minute returns from the UK stock exchange (Abhyankar, Copeland, and Wong 1995, 865) as well as price data for currencies, gold, and silver (Abhyankar, Copeland and Wong 1997, 2–3). Furthermore, the stock market crash in 1987 and other "market breaks" display a striking lack of proportionality between changes in information and changes in prices. Such breaks cannot be explained within the linear framework of the EMH in its semistrong form, which requires a proportional relationship (Cunningham 1994, 572, 594). It seems possible that prices sometimes fluctuate not only because of new external information or even irrationality among investors but also because of the internal dynamics governing the market (Mandelbrot and Hudson 2004, 21). Cunningham draws a dramatic conclusion from these findings: the presence of these nonlinear dependencies "eviscerates the random walk model" and renders the EMH "largely meaningless" (1994, 581).[21] Philosophy of science has made us familiar with the notorious difficulties in falsifying a hypothesis: one can always rescue a favorite notion by tinkering with the auxiliary hypotheses necessary to flesh it out.[22] One approach, called "noise theory," attempts to save the EMH framework by distinguishing between informational efficiency (the notion that the market reflects all information, including irrelevant or misleading or irrational factors) and fundamental efficiency (the notion that the market reflects all and only information relevant to the fundamental or intrinsic value of the security). The euphemism "noise" allows economists to acknowledge that securities markets are infected by trading on information unrelated to intrinsic value. These irrational behaviors—bandwagonism, speculative bubbles, the nonnegligible causal effects of the Super Bowl—get labeled noise, whereas the EMH framework remains in place. Yet Cunningham asserts that noise theory cannot explain persistent nonlinear structure because there may be other factors besides behavioral irrationality to explain deviations from fundamental values (1994, 565, 593).

Recall that the EMH implies a random walk, and a random walk implies unpredictability. The presence of nonlinear dependencies suffices to invalidate the random-walk model and thus render the EMH untenable, but such dependencies may still be useless for generating predictions of securities prices. Consider the following analogy: the fundamental principles of special relativity yield Einstein's equations for moving bodies, which in turn imply that it is impossible to go faster than the speed of light. If someone were to disprove one of Einstein's equations for moving bodies, the feat would create a serious problem for the fundamental principles of special relativity. But it would not necessarily enable faster-than-light travel. As noted in chapter 4, the evidence for low-dimensional chaos in financial data remains tenuous. But if a market process is chaotic then, even though it is not "efficient," there may be no way to exploit the nonlinear structure to earn a profit. Even supposing that the deterministic structure can be discovered, sensitive dependence may make it impossible to forecast accurately enough to be profitable (Abhyankar, Copeland, and Wong 1995).

Policy Implications of Challenges to the EMH. The EMH "explains" something that does not happen—markets behaving according to a random-walk model. Cunningham allows that the EMH may be useful to economists for understanding some types of behavior, or behavior that sometimes occurs, but he draws the pluralist conclusion that "it would be difficult to defend transplanting such a partial theory to the domain of public policy articulated by lawyers" (1994, 581). Just as policymakers should be cautious in relying on the "efficiency" of markets to evaluate legal rules, "they should be cautious about relying on nonlinear dynamics and chaos theory in doing so as well" (607). Market phenomena are complicated and probably not best approached with a single account.

In addition to this methodological lesson, Cunningham offers some substantive policy suggestions. For example, he invokes chaos theory in support of a new kind of "circuit-breaker" policy for interrupting market crashes. He suggests a circuit breaker that slows down the "intrinsic time" of a market by slowing trading when it gets too fast, in order to curb crashes as well as hyperefficiency's overexuberance (1994, 600–601).[23] Cunningham also notes that the EMH has been used to deny the need for investor protection in the form of mandatory disclosure regulations, on the supposition that all necessary information is already included in the price. In contrast, noise theory makes disclosure irrelevant by depicting investors as irrational (602). Cunningham makes use of nonlinear dynamics to cut through an analysis of market behavior that he views as "aridly binary": we do not have to choose between rational efficiency and irrational noise (608). Some disproportionate responses to external information may be the result of structural factors within the market's own dynamics; nonlinearity need not be irrational.

And what makes markets inefficient may be more than just noise. Non-linear dynamics and chaos theory, by casting doubt on the EMH, serve to support disclosure regulations that protect investors. In fact, Trig Smith has argued that despite some explanatory limitations, chaos theory "should remain as one of the guiding factors for the development of S.E.C. policy" (2000, 785). With his connections to urgent matters of policy, Cunningham demonstrates that the busting of dichotomies can be of more than merely conceptual interest.

Legal Remedies

Inspired by the economic conclusions of the "calculation argument" described earlier, some may be tempted to make a general argument against most any kind of social intervention. After all, according to Hayek, "the number of separate variables which in any particular social phenomenon will determine the result of a given change will as a rule be far too large for any human mind to master and manipulate them effectively" (1952, 42). Although Hayek's point depends on the sheer number of variables involved in social systems, Pierre Lemieux draws a similar conclusion from chaos theory: "Confronted with a dynamic and nonlinear system like society, the planner can predict neither its actual development nor the results of his interventions. . . . Intervening in social developments is as futile and dangerous as intervening in the weather" (1994, 25).

Yet some legal scholars have drawn far different implications from chaos theory about whether and how governments should act to reorder society. Some have invoked the chaotic character of human behavior as a reason why we cannot know how a particular regulatory intervention will work out, while leaving open the possibility of effective governmental action (see Downs 1995, 31; Roe 1996, 667). In examining banking regulations meant to ensure equal economic opportunity, for example, Vincent DiLorenzo takes the moderate view that government cannot achieve all desired social outcomes because much control is in the hands of corporate decision makers, yet he suggests that governmental intervention is warranted when private markets cannot achieve the desired social goals and must be "rechanneled" (2000, 54, 58). Before governmental intervention, banks' response to urban problems was primarily to rely on traditional lending criteria that treated all potential clients in a uniform way. But instead of the flexibility that a "strange attractor" lending pattern could provide, the free market was found to follow something more like a fixed-point attractor because of profit incentives and the desire for uniformity (111). Recall that the complicated, never-repeating patterns of strange attractors can represent the dynamics of a chaotic system, whereas nonchaotic systems may settle into simpler patterns or static points. I set aside until the next chapter the question of whether flexibility is a virtue and concentrate here on DiLorenzo's suggestion for a different approach to governmental regulation. According to him, chaos theory embraces unpredictability and instead seeks responsiveness, so legislative intervention modeled after a strange attractor can elicit socially responsible action, which the free market alone would not do, even though such action would be profitable (2000, 120). But in an earlier essay, DiLorenzo examined the issue of campaign finance reform from the perspective of chaos theory and reached a very different conclusion about the efficacy of regulatory intervention: "In the light of chaos theory any hopes of substantial changes in congressional decision-making appear groundless" (1994, 485). In this earlier work, DiLorenzo attempted a time series study of condominium legislation in New York State to identify all the different factors that affected legislation. This questionable use of quantitative analysis purportedly "demonstrates the first principle of chaos theory: There are many causes of legislative action, and no single cause constantly determines outcomes" (439). But the presence of many causes with time-varying strengths would actually make a low-dimensional chaotic attractor quite implausible.

Other legal scholars have made more convincing arguments for a reconceptualization of legislative intervention. Examining environmental law, for example, J. B. Ruhl urges us to distinguish between the question of whether a particular regulation is a good one and the question of whether we have a good overall system for responding to new environmental challenges. We cannot predict what new challenges will arise or what chaotic behaviors will manifest, so

> the best we can hope for is to gain some measure of control generally over how frequently and intensively those behaviors occur, and to harness their normatively positive effects on a system-wide basis. The first step in that direction is using the model to understand what is happening in the system; only then can we begin to ask whether the system is on the course we desire. The descriptive value of the dynamical systems model lies in its ability to assist us in knowing where and when we may need to attempt to adjust the trajectory; the analytical value of the model lies in its ability to instruct us in how to do so. (Ruhl 1996, 886; see also Young 1997)

This approach should remind us of the parameter-tuning model for economic systems discussed earlier.[24]

Michael Wise, an attorney at the Federal Trade Commission's Bureau of Economics, has written at length about the implications of chaos theory for antitrust regulations. As we have seen in the chapter 5, Wise invokes chaos theory as a new metaphor that can serve as an antidote to the influence of previous importations from physics. He believes that influence has led to a pernicious scientism because economics has substituted a physics-style technical analysis for substantive inquiry into evaluative questions about the distribution of wealth and power. "Claims that science is a uniquely authoritative justification for legal principle become claims about who should have power. While science has a better understanding than law does about the causes of metal fatigue and the biological significance of blood typing, it is doubtful that science has a better understanding of how people make contracts or of how they exercise power over each other through their economic and social institutions" (1995, 725). Wise's concern about this scientism manifests in pleas for policy goals beyond efficiency and in his criticism of economic discourse based on nineteenth-century physics: "Adopting the techniques of physics reinforced an individualist conception of value and the utilitarian understanding of social welfare"—that is, isolated economic actors "evoke" the particles of physics, and there was the promise of being able to measure the general welfare by adding up individuals (729; also see Sen 1987 for a critique of utilitarianism in mainstream welfare economics). According to Wise, standard neoclassical price theory sacrifices vital evaluative considerations in order to achieve mathematical clarity, and policy based on this model ignores factors such as justice and power "because trying to represent these issues prevents the model's mathematical methods from producing answers" (733). So how can chaos theory help secure the autonomy of evaluative inquiry in the face of the mathematization of economics?

Recall our discussion about the rhetorical power of the natural sciences to help motivate methodological change. Michael Wise was one of those who made use of the prestige of chaos theory to help persuade his readers to let go of a focus on stable equilibrium states and explore instead the dynamic evolution of institutional arrangements (see chap. 4). This methodological reorientation has profound implications for antitrust policy, according to Wise. After all, chaos theory has been used in theoretical models of "duopoly, advertising expenditures, and the theory of the firm. That is, it may prove important not only for grand metaphor, but also for the very applications where economics has intersected antitrust law" (1995, 759). The new borrowing from chaos theory provides new assessments of what is possible and new value-laden metaphors of what is the "natural," "optimal" state of a competitive economic system.

Wise cites Brock and Malliaris to the effect that "systems subject to chaos may fail to reach the usual Pareto optimal ideal even under very sympathetic assumptions. In the presence of chaos there is not even a theoretical basis for faith that a formal equilibrium system produces an ethically optimal outcome" (Wise 1995, 761). So economic analysis of law should focus on paths and rules that govern a system's evolution, comparing the effects of legal interventions on systemic dynamics rather than on the equilibrium points they may never reach. Wise argues that antitrust policy has adopted the goal of maximizing competition and has fixated on a criterion of evaluation that favors static equilibrium states. But chaos theory casts serious doubts on the assumptions guiding current policy: economic systems may never reach static equilibrium, such states may not be optimal even by the conventional Pareto measures, and it may be impossible to discern the exact effects of policies that seek equilibrium (761, 769).[25] Antitrust law based on a metaphor borrowed from chaos theory would not yield rules that determine outcomes of competitive processes, for "the likely impossibility of such rules is deterministic chaos's principal lesson." But it could support principles that deal with problems of change and growth (777).

A similar suggestion has also been made by the physicist David Ruelle, based on the tentative identification of sensitive dependence, if not low-dimensional chaos, in many time series data from economics and finance: "The complicated arrangements elaborated by political decision makers in search of a subtle optimal equilibrium may be doomed to failure. Accepting the idea that the future is chaotic and not foreseeable in detail, one should set up regulations and arrangements that are robust in the presence of unforeseen events" (1994, 28). So instead of crafting regulations that seek an impossible goal of dictating outcomes whose justice has never been established, Wise suggests that understanding nonlinear dynamics "might help design rules that maintained development within generally known and acceptable bounds, yet preserved the potential for change and accommodation of the unexpected" (1995, 769). Notice that Wise relies here on the value of adaptability, and we see this also when he suggests that constraints on free entry could be disfavored as preventing flexibility and freedom of action (770). We can see here that Wise relies on the value of flexibility to show how his alternative account could "supply a metaphorical foundation for evaluative criteria" (770). Wise is circumspect enough to recognize that policymakers are unlikely to abandon the clear old ideas "just because they are wrong," and he judiciously declines to speculate on the exact form of new rules for antitrust law based on the new metaphor (775). But he argues that the antidote provided by chaos theory can defamiliarize unquestioned justifications for existing policy and lead to "explorations in new, heretofore discouraged, directions" (776).

How then can chaos theory combat the foreclosure of evaluative debate by technical analysis? By undoing the presumption that neoclassical economic science has provided an unimpeachable goal and a surefire means to achieve it:

> The major lesson of deterministic chaos would be that policy choices cannot depend on outcomes predicted by scientific authority, for science has found that outcomes may be fundamentally unknowable. Rather, policy choices will have to depend more on ethical choices about the process.... Science cannot do the law's ethical work.... A new metaphorical foundation may make it more legitimate to argue that law must explicitly recognize the importance of reliable expectations of fair treatment, and must recognize and deal with substantial variations in individual economic circumstances, for its system of rules to generate a stable yet supple dynamic path. (Wise 1995, 774–76)

The defamiliarizing effect of the metaphorical borrowing from chaos theory can sever the implicit "ought" from the explicit "is" of conventional economic theory, making the leap from description to evaluation a topic for open discussion. And this is a valuable service, for as John Dupré writes, "what *should be* the goals of economic systems is one of the most fundamental problems of the discipline" (1993, 213). When technical research (and the prestige borrowed from physics) has helped to allow economics to close off debate about this fundamental evaluative issue, an antidote borrowed from more recent physics may be just the thing to reopen it. If chaos theory cannot provide an exact blueprint for effective governmental regulation, at least it can open up a space for debate about crucial ethical considerations.

To conclude this section on chaos theory and intervention, I note that literary scholars have also commented on the problem of whether and how one should act in the face of the uncertainties of nonlinear systems. Most notably, Peter Mackey argues against Clive Hart, whose pessimistic view of *Ulysses* sees it as full of failed hopes, unexpected outcomes, and ultimately nothing much happening at all. Mackey contends that the protagonist Leopold Bloom chooses to act even in the face of uncontrollable chance events, and chaos theory helps us to see the heroism of small hopeful efforts in an uncertain world (1999, 166). Indeed, *Ulysses* holds out a hope for meaningful interventions at the scale of individual human lives because a small change in a fated yet unpredictable world can still make a difference. As Mackey writes, "chaotic as his life may seem, ignorant as he may be before the future, Bloom proves he has the resolve to become a 'conscious reactor against the void of incertitude'" (201–2).[26] In our everyday lives, as well as in the realm of governmental regulations, chaos theory need not bring a counsel of despair nor remain silent on questions of value. Instead, nonlinear dynamics can be profitably read as an inspiration to thoughtful and modest action even when there is no hope of exactly predictable outcomes.

8 Beautiful Chaos?

Knowledge borrowed from the natural sciences can inform our inquiry into questions of value, even though it cannot settle such questions on its own. But theoretical discussions of values sometimes marginalize aesthetics as being utterly separate from, or irrelevant for, ethical and political questions. In this chapter I examine two of the most fully developed attempts to use chaos theory in aesthetic arguments about literature: the work of Alexander Argyros and Harriet Hawkins. First, I examine Argyros's attempts to use science to ground universal norms in both aesthetics and politics. Then I turn to Hawkins's articulation of fractal structure as the basis for aesthetic evaluation. As we will see, the deployment by Argyros and Hawkins of such concepts as informational complexity and fractal structure touches not only on issues of beauty, but also on questions of goodness and truth.

Argyros and the Value of Complexity

Using Science to Ground Universal Norms
In his book *A Blessed Rage for Order*, Alexander Argyros undertakes the bold project of grounding aesthetic and even political norms on contemporary science. He holds out the possibility that "recent developments in science may be confirming, not contesting, certain traditional humanistic

ideas" such as "order, identity, truth, humanity, ethics and aesthetics" (1991, 91–94).[1] Argyros characterizes his own project as "delightfully eccentric" (7), wild and irresponsible speculation (7), and "petulant at times" (348). Yet the seriousness and detail with which he formulates his arguments invite careful critical scrutiny. How can it be that science confirms artistic and ethical ideals? For Argyros, science can accomplish this task because nature itself has "preferred strategies" that set the standard. At times, he makes the moderate suggestion that the natural world does not impose binding norms but merely makes it possible to formulate evaluations "along a scale of greater or lesser likelihood to engender productive effects" (209). In this, he makes an argument not far from the pragmatist position defended in chapter 7. But at other times he asserts that the standards of the natural world are "the only standards that matter" (7) and that the "interpretive community which selects for such humanistic concepts" is "the ten- or twenty-billion-year history of our evolution into human beings" (95), as if the cosmos itself is an interpretive community that can make evaluative choices. And Argyros sometimes attempts to read ethical conclusions directly off scientific facts: for example, when he speculates that because the universe generates novelty by having parts of itself loop back reflexively, "our chief responsibility to the universe" is "to be an instrument of its introspection and evolution" (115).

One of the ways Argyros uses chaos theory in his project is as a source of metaphors for a theory of cultural universals. For him, "the existence of cultural universals, or cross-cultural tendencies to organize the world in similar ways, could serve as a kind of foundation for aesthetic decision making that is neither parochial nor rigidly normative" (1991, 210). These universals arise from the innate predispositions that result from our evolutionary heritage and impose physical and biological constraints on concept formation (211, 219). Among these putative universals are "belief in a deity," "agriculture," "face-to-face sexual intercourse," and a negative value assigned to "smallness of spirit, and snakes" (214–15). These cultural universals would provide us a kind of normative aesthetics, because they are ignored at the risk of irrelevance (346). Argyros spells out their role using some metaphors from chaos theory: "*God* and *beauty* are cross-cultural attractors," and the understandings of different cultures nonetheless trace similar attractors in their "mental phase space," which allows intercultural communication (297). A potential objection to his theory is likewise addressed using terms borrowed from the study of chaos: "the theory I am presenting does, indeed, appear to sanction the status quo. For example, if such concepts as city, love, and beauty are universal attractors, then why should not genocide, racism, and patriarchy also be similar basins of attraction? And if the latter are universal, how can there be reasonable hope for change? In a way, I plead guilty to these charges" (301). But he goes on to claim that cultural universals are not eternal and are "actually essential to the emergence of new cultural possibilities" because innovation needs some stability (302). Genetic constraints are like attractors, and cultural innovations push the system into new patterns of organization.

This lofty project faces a number of problems, and the metaphorical use of chaos theory may not be the worst of them. Although Argyros expects that "a cross-cultural sample would confirm the near universality" (1991, 297) of basic experiential concepts, the data are certainly not conclusive at this time. Despite his assertion of universal negative valuation, some people like snakes. And although he suspects that a weighted list of the statistically prevalent themes across cultures would be "a stunningly accurate tool with which to make heuristic guesses of a work's potential importance," (223) his suggested list of themes includes sports, cleanliness, dancing, modesty concerning natural functions, property rights, surgery, and weather control (221). I am not sure how many of these we can find in *Paradise Lost*. For the purposes of this study, however, perhaps the most relevant problems are those of fit and distinctiveness: Argyros seems to have simply helped himself to the terminology of chaos theory without paying attention to the difference between a strange attractor and its basin, or to the fact that the terminology of attractors adds nothing to his argument except a faint sheen of scientific prestige.

One of the cultural universals whose value Argyros most eagerly seeks to defend is traditional narrative structure. Traditional narrative has suffered from a postmodern assault on conventional literary techniques, especially Lyotard's (1984) declaration that our culture has exhausted the grand narratives of progress and freedom that once gave it meaning. One of Argyros's main weapons in his fight is chaos theory: "the sciences of chaos and of dissipative systems may offer the theoretical muscle to allow us to affirm the desirability, legitimacy, necessity, and morality of narrative in general and grand narrative in particular" (1991, 315). I examine the tangled relationship between chaos theory and postmodernism much more fully in the next chapter, but for now let us focus on Argyros's use of science to defend the value of traditional narrative.

In pursuit of his goal, Argyros proposes a sociobiological theory of literature whereby it "enhances the survival potential of humans" by allowing a culture to represent and evaluate hypotheses about the world and its future choices (1991, 196–97). On his account, narrative is "one of the most remarkable and desirable inventions of biological evolution" (309) because it serves as an especially robust kind of "information-processing strategy" (315). Unfortunately, Argyros relies on discredited sociobiological arguments to reach his normative conclusions.[2] His argument has three stages. First, "the universality of narrative implies that it reflects an underlying neural substrate or a set of epigenetic rules predisposing human beings to organizing experience in a narrative manner"; second, "we can assume that narrative is a good match to significant features of our environment" conceived of as causal relationships; and third, "we can hypothesize that narrative is probably self-similar to those innovative processes in nature whose engine is chaos," mirroring the way natural systems balance conservatism and change (316).

In the first step, we see again his sweeping assertion of cross-cultural universals, followed by Argyros simply helping himself to a biological basis for each alleged cross-cultural feature. In the second step of the argument, an optimistic adaptationism then takes over, asserting that any biological feature of a species must be well suited to its environment. The result is a well-adapted feature of human nature which we ignore at our peril. The normative import derives from the suggestion that if we act in ways that conflict with the adaptive patterns of narrative, we will encounter an environment that will deal harshly with us. As mentioned in the previous chapter, considerations of human nature are not irrelevant for evaluative inquiry, but such considerations must be based on more substantial evidence than Argyros provides.

Chaos theory is invoked in the final stage of this argument. For Argyros, "traditional narratives can be viewed as chaotic laminar systems, rivers characterized by an overall vector, the plot itself, composed of areas of local turbulence" (1991, 318). Already we can see a serious problem because layers of flow in laminar systems move as separate sheets, whereas flow in chaotic systems necessarily involves mixing. A "chaotic laminar system" is a contradiction in terms. But the problems go much deeper and rest on a fundamental confusion about strange attractors. Witness: "Using the insights offered by chaos theory, it is tempting to speculate that traditional narratives are, in fact, far-from-temporal-equilibrium dynamical systems capable of generating global order simultaneously with local randomness. The remarkable feature of chaotic systems is their tendency to settle into perdurable patterns, chaotic attractors, that are nevertheless highly sensitive to external fluctuations and initial conditions" (319). Argyros wants to use strange attractors as a metaphor for narratives that reconcile stability and change. But he fails to recognize that although strange attractors are often structurally stable, and not sensitive to fluctuations in parameters, they are also generally very robust with regard to initial conditions. Outside perturbations or changed initial conditions will alter the detailed *trajectory* of a chaotic system but will still yield the same strange attractor so long as the trajectory remains within that attractor's basin. So chaotic behavior displays sensitive dependence on initial conditions, but the strange attractor itself displays a marked *insensitivity* to initial conditions. Argyros wants chaos theory to offer systems that exhibit both patterns and creativity, but that is the much more speculative province of self-organizing systems and complexity theory. His attempt to move from facts to values founders on a failure of metaphorical fit.

Does Nature Prefer Complexity?

Argyros seeks to use chaos theory to support evaluative conclusions about the aesthetic importance of traditional narrative and also the superiority of liberal democracy. At several points in his book, Argyros assumes a preferred status for complexity. He proposes that "the most robust and complex things in the universe, including human beings and their theories, works of art, and social structures, are best understood neither as Platonic essences nor as random processes, but as chaotic systems" (1991, 6). Leaving aside the fact that this is a difficult claim to substantiate, I focus instead on the presumption that complexity is somehow preferable in a way that can ground aesthetic and political choices.

At crucial points in his argument, Argyros simply asserts that natural processes tend to bring forth new and complex structures over large time-scales. In his assertion that nature itself seeks to maximize complexity we find an example of the rhetorical strategy identified by John Lyne, which "finds in nature itself the definition of long-term, higher purposes" and thus "reduces the distance between is and ought and helps us manage the tension between them" (Lyne 1990, 42). For example, at a crucial juncture Argyros invokes the popular science writing of Paul Davies, who speculates that the universe might have a "preferred direction of evolution" leading to the step-by-step unfolding of systems with increasing complexity (Argyros 1991, 258; see also 150, 325).[3] In this view, complexity is "inherently valuable" because it increases the number of options from which an organism can choose (141) and also because it makes an organism's life "more intense" (149). But Argyros fails to consider the substantial criticisms of this view: that it mistakenly imports teleology into neo-Darwinism, that it conflicts with the historical record that shows some features of organisms becoming less complex, and that it promotes statistical artifacts to the level of cosmological principles.[4] Argyros even admits that more complex animals may be more vulnerable, making his justification for the value of complexity vacuous: if more complex animals thrive, this proves their greater efficiency and robustness, and so complexity is superior; if they die off, or are perpetually outnumbered by beetles and blue-green algae, then their life is so much richer that complexity is again proven superior.

Stephen Jay Gould advanced several well-known criticisms of the view of evolution as a triumphant march toward increased complexity (see, e.g., Gould 1994). Argyros briefly acknowledges this disagreement in a footnote that cryptically asserts that "most evolutionary views, however, differ from deconstruction in positing some form of nonconstructivist continuum between the prehuman and human worlds" (1991, 351). This comment repeats his assertion that deconstruction is incompatible with science, but it does nothing to answer the criticism that Argyros has based his normative conclusion on the questionable assumption that the universe *prefers* increasing complexity.

We find this normative leap expressed when Argyros summarizes his argument as follows: "If the universe is, indeed, a society of chaotic, self-similar layers, then it appears that everything in nature...works best when it resembles a chaotic attractor" (1991, 331). But the antecedent condition is in fact false: the universe is not made up entirely of "chaotic, self-similar layers." And Argyros contradicts this point himself by asserting that nature prefers systems that have regular patterns (250). But more importantly for the examination of normative claims deriving from chaos theory, even if the universe is the way Argyros wants it to be, his conclusion does not follow. Even if the universe were entirely made up of systems with a certain character, this arrangement would not establish that systems of that character "work best" unless the universe as a whole operates to select for systems that "work well." Argyros may argue that in the long run, only robust systems will persist, but this line of reasoning assumes that new systems are not continually generated. Of course, if by "working well" we mean something like "obey the principle of least action" (or some other physical law), then all physical systems necessarily work well. Argyros seeks to elevate a principle of increasing complexity to the status of a long-run statistical law, like the second law of thermodynamics, but there is no justification for such a move.

Even if chaotic systems did work best, in the sense of generating robust and complex behavior, this scientific result could not ground a substantive ethics or aesthetics without considerable additional work. Argyros remains unable to answer the question, "Why are more complex systems better than simpler ones?" Perhaps science will one day establish that social structures that instantiate chaotic attractors produce novel ideas and resist external disruptions better than other social structures. One still must ground a preference for novelty and robustness, and facts about the universe cannot immediately provide such a ground. Argyros merely suggests that arranging society or interpreting texts in ways that inhibit chaotic processes would have consequences generally thought to be unpleasant, and although his suggestion may be borne out by further investigation, it is hardly conclusive as it stands.

Looking more specifically at literature, we find that Argyros (unsurprisingly) holds a conventional notion of literary merit based on formal unity: "The best literary works orchestrate their component structures so that out of their harmonic interaction can emerge an experience of heightened unity and identity" (1991, 196). Yet he identifies complexity as the very basis of beauty: "I think that this is precisely what constitutes the beautiful—the unpredictable and discontinuous emergence of higher levels of systemic complexity" (287). Other writers share this scale of value, as we see when Michael Vanden Heuvel praises the Wooster Group's work *Nyatt School* as "infinitely more complex" than T. S. Eliot's play *The Cocktail Party*, which it partially contains and comments upon (1993, 265). But the use of complexity to establish norms of aesthetic merit faces special problems. It is one thing to suggest that measures of informational complexity useful in chaos theory might provide useful analogues for characterizing literary texts. It is another thing to claim that complexity yields an objective measure of literary merit. Objectivity in aesthetic judgment is a possible achievement, but an achievement to be gained by the hard work of building interpretive communities with agreed-upon standards and widely accepted goals for literary production. In the absence of such agreement, the borrowing of scientific measures runs the risk of merely reinscribing conventional distinctions between "high" and "low" art, and of giving this invidious distinction the gloss of scientific objectivity borrowed but not earned. The fact that we can build bridges from facts to values does not enable us simply to hop across the gap unaided by satisfactory arguments.

Adaptability as a Value

In further defending his normative project, Argyros asks the question, "What form of contemporary social organization would be most likely to encourage our culture to make the new world as humane as possible?" (1991, 328). This question is indeed important—crucially important—and Argyros is to be applauded for raising it. Argyros goes on to speculate that chaotic behavior, rather than randomness or strict order, "appears to be the one nature tends to choose in all its integrative levels when survival depends on flexibility and innovation," and then concludes that we should choose as nature does (247).

> The universe values managed creativity above all else, and a society that is able to manufacture more innovation will be more likely to survive in the long run than a rigidly authoritarian one. I believe that a society whose most revered product is controlled novelty will tend to organize itself in a flexible and frequently tangled hierarchy because such social systems have a higher probability of success than those, such as traditional patriarchy or institutionalized racism, that simply exclude from the evolutionary arena vast amounts of potential human creativity. (330)

Even those of us with sympathy for his liberal conclusions may be suspicious of the ease with which Argyros simply reads efficiency in the information economy as the key to moral values. In supporting his case for a humanistic, tolerant society with a mixed economy, he elevates the principles of liberal democracy to the status of deliverances from nature itself in the passage I quoted earlier: "Ultimately, I believe that chaos offers a bracing vision of political normativity. If the universe is, indeed, a society of chaotic, self-similar layers, then it appears that everything in nature, from prebiotic dissipative systems, to the ecosystem of a river, to the organization of a primitive nervous system, to the dynamical flow of a human brain, to the shape of a kinship group, a city, a nation, or a world works best when it resembles a chaotic attractor" (331). Of course, if the universe really is a community of chaotic systems, then it becomes pointless to assert that things work best when they resemble chaotic systems. As we will see in some detail in the next section, Argyros falls into many of the traps that await those who collapse the evaluative into the factual; in this case it is the temptation to assert that whatever is, is good.

The value that Argyros places on adaptability appears to be shared by many of those exploring connections between nonlinear dynamics and the law. For example, Vincent DiLorenzo and J. B. Ruhl argue that legislative regimes modeled on chaotic systems are preferable because they are more adaptable. DiLorenzo proposes that a legislative system that allows variation in its responses to different relationships between corporations and the community is "a potential strange attractor" that exhibits a kind of sensitive dependence (2000, 83). He argues that "there are advantages to legislative schemes that permit chaotic turbulence. Such legislated schemes generate greater responsiveness—responsiveness by corporations to community needs and demands" (82). As an example of such a scheme, he offers the 1977 Community Reinvestment Act (CRA), which recognized that government alone could not solve urban decay. The CRA and attendant legislation "reconfirmed the existence of a potential strange attractor—rejection of the imposition of a predictable outcome" in favor of a flexible set of factors (91). Similarly, Ruhl argues that legislative reform needs to look at the qualities of robustness or sustainability that allow systems to respond well to new challenges, and one of these key qualities is adaptability, the ability to respond to challenges (1996, 887). Stasis is not sustainable, so "strange attractors are a key ingredient of robust systems, and thus some level of chaotic behavior is necessary to maximize system sustainability" (892). In Ruhl's view, a "broad social preference for predictability" has motivated Congress to focus almost exclusively on regulatory approaches that attempt to micromanage reform by fixating on the simplification of specific rules (1996, 912). A perspective informed by chaos theory, however, would pay more attention to the question of what legislative system enables adaptability.

Economist J. Barkley Rosser Jr. notes an analogy with the notion of a "stability-resilience" trade-off in ecology. Rosser draws on the work of noted ecologist C. S. Holling in asserting that "populations or systems that are highly stable may be vulnerable to discontinuous and catastrophic collapse whereas unstable systems may be more resilient. The latter adapt more readily, if more chaotically, to environmental changes, whereas the former tend to rigidly resist change until they collapse or fall apart" (Rosser 1991, 322). And we find the idea that chaotic behavior promotes a desirable adaptability in the work of a wide variety of other legal scholars as well. Glenn Harlan Reynolds argues that it may be a good thing that the Supreme Court fluctuates in its positions without ever reaching a final answer, because "the 'chaotic' nature of the judicial system may mean that stagnation through special-interest domination is unlikely over the long term, as periodic shifts by the Supreme Court lead to the periodic need to renegotiate political/economic alliances. The payoff from this could be significant in maintaining political and economic flexibility" (1991, 115). He goes on to make explicit use of the thick descriptive term "health" by arguing that courts are like living things, and that chaos is healthy in biology, so therefore legal rigidity is bad (116). Robert Scott states that "the phenomenon of patterns formed by unpredictable and irregular human behaviors is a reality that should give us comfort in accepting the inevitability of paradox in law. Do not despair because law has fundamental contradictions. It is the very tension whose resolution we seek that keeps our legal system in a dynamic state of continuous renewal and repair. It is the dynamic of the Justice Paradox that keeps our legal system alive" (1993, 350). Gordon Slethaug draws a similar moral about adaptability in his discussion of literature and "systems theory": "All systems must change in order to continue, for what makes a system strong at one time may well account for its weakness and cessation in another" (2000, 46). And Williams and Arrigo use a chaos theory metaphor for society as a whole, claiming that strange attractors allow a society a healthy flexibility, whereas stasis is death (2002, 144–47).

Not everyone agrees, however, that being adaptable is such a good thing. Barondes mocks as "senseless" the assertion by Scott (and Reynolds) that chaotic oscillations are helpful in a legal system, arguing instead "that a judicial system produces varying results when there is no change in the relevant parameters does not necessarily imply that it will produce desirable changes in results when there is a significant change in a relevant parameter. The best that can be said of such haphazard results is that the results occasionally will be correct" (1995, 168). But Barondes has missed the crucial point that the unpredictable behavior of a chaotic system is not "haphazard." Those offering chaos as a model for healthy flexibility need not assert that a chaotic system will always change in the right direction, only that such a system will be less likely to experience catastrophic failure when confronted by a changed environment.

Adaptability can certainly serve as a bridge from fact to value. But is it always good to be adaptable? One way to construct the bridge rests on the notion that adaptable systems are better able to survive. A famous advertisement for Cadillacs once declared: "That which deserves to live—lives," but even Friedrich Hayek—who sought to construct evolutionary analogues for economics—recognized that simply surviving evolution does not grant moral value (Hodgson 1994, 435; for the advertisement, see McManus [1915] 1959). Sometimes things need to die, and sometimes what survives is not the best, nor even particularly good. An important feature of some nonlinear systems is the "locking-in" of suboptimal entities as a result of path dependence. What survives can often depend on utterly accidental features of the initial situation, so existing economic and political solutions may not be the best ones (Roe 1996, 641–42).

We must ask, what is so great about responsiveness? At a time when corporations insist that their workforce be "flexible" so that the company can jettison workers with a minimum of bother, do we risk simply validating hypercompetitive managerial jargon by reading it back into nature? I am not against adaptability, of course. It can be a useful way to build an argumentative bridge from scientific results to evaluative discourse. But it must not short-circuit the inquiry process by simply baptizing certain preferred arrangements as "adaptive" or "flexible." As Dewey counsels us, we must always ask whether adaptability (or any other thing we desire) is actually valuable in a particular case. Should universities be arranged without discipline-based departments, so that they can adapt to changing conditions in the intellectual marketplace? Should educational institutions be flexible enough to respond quickly to the latest career fad that strikes students' fancy, or the most current pressing needs of corporate job recruiters? Ruhl points out that a sustainable legal system requires more than just adaptability; it requires a certain type of stability and simplicity as well. For Ruhl, "stability" means that relationships between agents remain similar over time, "simplicity" means that relationships between agents are easily determinable, and "adaptability" means the ability to respond to challenges. But stability and simplicity do not require rigid rules, and adaptability does not require an immediate or direct response to each emergency (1996, 887–88).

For many of those who borrow from chaos theory, the values of complexity and flexibility constitute articles of faith. But whereas chaotic systems provide metaphors of a more humane social order in some contexts, some writers view strange attractors as less hopeful images. For Ira Livingston, chaos theory excites a mixture of pleasure and suspicion, together with "grumpiness about hype." From this conflicted position, he seeks to "compensate for the widespread use" of such terms as "chaos" and "nonlinearity" as "progressive or even messianically liberatory" (1997, 3). And we have seen the reaction of mathematician René Thom against what he sees as the cultural celebration of randomness. Thom recognizes that opposition to determinism may be motivated by resistance to the oppressive power of technocracy, but he notes that technical experts wield their power in areas in which risk and chance hold sway. So progress in shrinking the realm of randomness actually reduces the power of the expert and contributes to the project of liberation (Thom 1983a, 83). These dissenting voices sound an important cautionary note.

Chaos theory can provide a valuable metaphorical resource for those who seek to promote a vision characterized by the evaluatively charged features of complexity and flexibility. But the bridge from facts to values can be a shaky one when it is too easily assumed that "complex" and "flexible" are good things to be.

Hawkins and the Fractal Aesthetic

Hawkins's main point is that chaos theory establishes and explains the special character of great works of classic literature. She sees two main features of chaotic systems as the keys to understanding literary greatness: the intertwining of order and disorder, and the self-similar structure of fractals. In what follows, I examine these two features to uncover where they prove helpful in constructing evaluative arguments about literature as well as where they are used less successfully. Allow me to remark upon how poignant it is that Hawkins feels she must defend the works of Milton and Shakespeare against a perceived attack on their worth by arguing that their beauty and relevance are demonstrated by contemporary science.[5]

Order and Disorder as a Universal Feature of Art

Hawkins argues that the interactions between order and disorder create the beauty and power of works such as Milton's *Paradise Lost*. Her discussion of this interplay begins with an admirable statement of the challenge in moving from scientific description to aesthetic evaluation: "As a scientific concept, deterministic chaos cannot conceivably be described in moral terms. Its frame of reference is mundane, not metaphysical. But it has obvious affinities with ancient as well as modern philosophical and theological metaphors involving synergistic interactions, as well as oppositions, between order and disorder" (1995, 4). She goes on to explicate her observations of these affinities (e.g., 28ff.), and her close reading of *Paradise Lost* is made richer by its careful attention to the intimate interaction between order and chaos in the text. Unfortunately, the reading suffers from the systematic conflation of the terms "chaos," "turbulence," and "disorder," weakening her argument by the overly flexible use of the scientific terminology.

According to Hawkins, chaotic dynamics and its metaphorical counterparts in literature "may prove closer to what most people find aesthetically fascinating and experientially relevant" than rigid theories about what people ought to like (1995, 164). This point seems considerably stronger, for she describes quite well the pleasures of controlled chaos within a work of art—pleasures that relieve us from the boredom of unrelieved stability yet do not overwhelm us with the threat of unconstrained randomness. As she puts it, "a combination of foreknowledge (determinism) and uncertainty within a work of art is far more engrossing and evocative of thrills and chills than pure predictability or surprise" (43). Indeed, John Dewey's *Art as Experience* describes the role of "continual variation" and the charm that a "touch of disorder" can bring: "the live creature demands order in his living but he also demands novelty. Confusion is displeasing but so is ennui" (1934, 167; see also Morin 1983, 29). But Hawkins goes on to make a much stronger claim about the generality of this literary metaphor: "Deterministic chaos is the rule in art: however dire the events portrayed may be, there is on the part of the audience a clear recognition of overall, preordained order in the fictional genre" (1995, 44). This universalizing tendency is examined in what follows, but in passing let us note a missed opportunity. One wishes that Hawkins would have seen that genre conventions function analogous to a strange attractor: providing a familiar and reliable large-scale pattern whose details cannot be precisely predicted. Unfortunately, Hawkins could not make use of this metaphor because she misunderstood strange attractors to be outside elements that interfere and bring disorder.

Despite all the ways in which literature can and does work to reinforce existing order, Hawkins claims that characters rebelling against order "are generally attractive to audiences *whatever* vices may be associated with them, *in so far* as they are associated with freedom, with the right to challenge, to defy, to disobey orders" (1995, 51). And she challenges us to find an exception. Here is an empirical claim about what is attractive to audiences, and the sources of pleasure in art. It is even possible that this argument may be a way to connect facts about the mundane requirements of sustaining social arrangements with some general claims about one of the sources of artistic value. Such an argument would have to address the fact that some audiences, especially those with a great deal of formal education, may actually prefer a high degree of randomness to works that display a balance (see, e.g., Short 1991, 348).

Hawkins, however, goes so far as to claim that chaos theory establishes the universality of some of these sources of artistic value. Again, she begins modestly, admitting that some aesthetic values are restricted to certain contexts but speculating that "the transcultural and transhistorical appeal of certain works of art suggests that there may also be some common ground" (1995, xi). She then brings chaos theory to bear on the question of whether literary merit depends on historical context, claiming that this new science "allows for a new way of discussing time-transcending forms and text-transcending relevance" (19). Later she links this issue of the context dependence of beauty to a claim that chaos theory "may ultimately doom to extinction" the rejection of universal truth because it "makes strong claims about its universality" (36). She cites a number of such universality claims made by scientists in Gleick's book but fails to notice that the universality being claimed is of a specific and circumscribed variety. For example, Mitchell Feigenbaum proved that for any one-dimensional equation with a certain mathematical feature, the transition to chaos will occur according to a specifiable pattern of "period-doubling." It doesn't matter what the real system is made of because the period-doubling pattern is "universal" for an entire class of equations. But this universality is hardly the kind that can establish transculturally universal forms of art.[6] After all, every decimal expression for a multiple of ten ends in the digit zero. This point is true universally, but it is hardly the kind of universality denied by those who see aesthetic or even ethical norms as culturally relative. The hard work of establishing transculturally valid generalizations about aesthetic experience, and then arguing that works that are valued prove genuinely valuable, cannot be circumvented by an easy appeal to mathematical universality, however convenient that might be.

Fractal Beauty

The popularity of chaos theory in the 1990s received a significant boost from the availability of intricate and brightly colored images of fractals generated by computers. As I discussed in chapter 1, these striking visual images have some connection with chaos theory because chaotic attractors typically display a fractal structure. Literary scholars who borrow from chaos theory make use of fractal imagery in a number of different ways, starting with the dust jackets of their books. Some are content to note precursors and anticipations of contemporary science, as when Ira Livingston cites Blake's view of the primacy of vortices in the physical world as well as in aesthetics as akin to Peirce's hypothesis of matter as composed of vortices of vortices ad infinitum (1997, 80). Other critics are explicit in their evaluative use of fractals, as, for example, Daniel Palumbo in his work on Asimov and Herbert, in which he attempts to defend the literary merit of their epic science fiction by detailing the fractal structure of their plots (2002). And the novelist John Barth praises fractal structure when he outlines his "principle of metaphoric means" in one of his essays on chaos theory. This general evaluative principle of skillful craft commends "the writer's investiture of as many aspects of the text as possible with emblematic significance, until not just the 'form' but the plot, the narrative viewpoint and process, the tone, the choreography, in some cases even the text itself, the fact of the artifact—all become signs of its sense (the way Dante's *terza rima* reflects, on the microscale, the macroscale of his three-part *Commedia*, itself a reflection of his tripartite eschatology)" (Barth 1995, 341).

Despite the desire to establish precursors and anticipations, literary scholars have missed a number of opportunities to notice some remarkable examples of metaphorical fractal structure. Linda Hughes and Michael Lund note that in Conrad's *Lord Jim: A Sketch*, "each part enacts in miniature the entire novel," but they never call this aspect self-similarity (1991, 175). A standard guide to *Ulysses*, Harry Blamires's *The Bloomsday Book*, notes that the first three chapters of the novel present the seed of the rest of the work by way of "microcosmic patterning" (1976, 9); that the "Wandering Rocks" chapter is "a miniature of the whole" and "a small-scale labyrinth" (94); that the opening section of the "Sirens" chapter is an "overture" that "lays before us, in concise form, many of the themes" of the chapter to come (108); and that the sections of the "Oxen of the Sun" chapter each relate to an earlier chapter (152). Yet Mackey's excellent work on Joyce and chaos theory (1999) fails to mention any examples of fractal structure. And generally I have encountered little mention of self-similarity in Hegel's philosophical system, even though Hegel himself described it as a "circle of circles" ([1830] 1975, 20), nor in the Hebrew Bible, even though Torah scholars have said that each part of it contains the whole (Ellis 1997).

As part of her project of grounding aesthetic norms in scientific findings, Hawkins makes use of the popular notion that self-similar fractal shapes are more beautiful than "linear" Euclidean forms. As evidence for this, she puts forward the fact that tourists from all over the world line up to purchase replicas of ornate buildings such as St. Basil's in Moscow because they find its fractal-like flourishes appealing (1995, 163). She also cites Mandelbrot and others who scorn the stark cubes of high modernism while praising as more satisfying the supposedly organic and self-similar curlicues of the baroque (56). In fact, she asserts that by means of his work on fractals, "Mandelbrot conclusively demonstrated that deterministic chaos in mathematics (as in nature) is aesthetically beautiful" (81). And Hawkins has company in her praise of the supposed beauty of fractals. Indeed, as early as 1986 an exhibit of fractal images at the Goethe Institute in New York was described by artist Gary Indiana as "gorgeous, something between microphotography and spin paintings.... They look like beautiful natural things: jewels, sunspots, snowflakes" (1986, 79).[7]

Hawkins then translates claims about the purported visual beauty of fractal images into claims about art in general and literature in particular. She cites with approval the suggestion by John Briggs that art "has always been fractal. The science of chaos is helping to newly define an aesthetic that has always lain beneath the changing artistic ideas of different periods, cultures and schools" (Hawkins 1995, 83). In support of this fractal aesthetic, Hawkins claims that throughout great works of art, including jazz music and Shakespeare's writing, "comparable symmetries and irregularities are replicated on every conceivable scale" (83; see also Gleick 1987, 117). And again, Hawkins hardly stands alone in her use of fractals as a guide to literary evaluation. Alexander Argyros asserts that "when we read a work of literature that we are tempted to describe as beautiful, I suspect that, at the very least, the work is a self-similar system, displaying similar patterns at different levels of description" (1991, 287).[8]

Hawkins sometimes makes the relatively modest claim that fractals offer a useful tool for appreciating a work of art. For instance, in her discussion of the theological, literary, and other contexts of William Blake's poem "The Tyger," Hawkins asserts that "all these frames of reference interact on every scale (linguistic, conceptual, symbolic, psychological, emotional) throughout the poem. Exactly the same holds true of the nonlinear geometry of nature as described in chaos theory: it is likewise operative, and contains a comparable complexity of informational content and detail, on every scale" (1995, 7).[9] But, as we have seen, she also puts forward a much grander theory of fractal aesthetics. What should we make of this evaluative use of borrowed knowledge?

The Ugliness of Fractals

A number of problems plague the fractal aesthetic, from simple errors to unwarranted assumptions to troubling omissions. Many attempts to use the concept of fractals have resulted in mistakes that fatally flaw their metaphorical borrowing, for example, when writers view each and every instance of plain old similarity (or mirroring, or tessellation, or hierarchical structure) as a case of "self-similarity" (see also Demastes 1998, 100; Livingston 1997, 51, 61, 81). Although not suffering from a lack of fit, some of Hawkins's examples of fractal structure nonetheless display little in the way of distinctiveness. For example, she locates one fractal aspect of Shakespeare's *The Tempest* in the fact that many characters in the play aspire to rule and that many characters are held in bondage of some kind (83). But these illustrations are hardly replication at *every* conceivable scale—one can conceive of mollusks and microorganisms being held in bondage, after all. Even granting that the claim of "every" scale is overblown, do we not find widespread "replications" even in comparatively meager forms of art? Consider a poorly made pornographic movie, which might well display comparable patterns involving rulership and bondage replicated over and over again.[10]

Yet the attempt to characterize beauty as fractal faces a problem more serious than just the occasional metaphorical mismatch. For what if fractals are not beautiful at all but instead ugly? I, for one, find the Mandelbrot set hideous. Detailed video depictions of this mathematical object that zoom in to reveal its endless hierarchies of detail make me feel nauseous, as if I am hovering above a garishly colored squashed bug with jagged lightning tendrils that dissolve into infinities of paisley when I move closer to try to grasp them. Perfectly ordinary-seeming spirals vomit themselves forth into subspirals made of jagged hooks and whirling eddies of endless airbrushed cloisonné. It makes me queasy because there is no ground, no place to set my feet and locate myself in relation to this object. Fractals are, in the technical sense, sublime. Precisely because they lack any intrinsic scale, they deny any reference to the scale of the human figure viewing them and precipitate one into a dizzying disorientation not so much pleasurable as merely trippy.

This little tirade can only serve to prove that the universality of fractal beauty requires further argumentation, especially in light of the profusion of contrary evidence. Certainly Hawkins can cite a number of attractive buildings with self-similar decorations. But a little reflection reveals that there is plenty of tourist appeal in structures entirely devoid of fractal adornment. Consider the pyramids of Giza—about as Euclidean as you get. Many other monuments around the world display beauty and popularity that fly in the face of any claim that humans have a natural affinity for fractal shapes. The Parthenon's facade, the Great Wall of China, and Stonehenge all demonstrate our fascination and appreciation for order, linearity, and predictability. And we can effortlessly multiply examples of cultural and natural objects whose beauty suffers not one bit for lack of fractal structure: Michelangelo's *David*, the paintings of Mondrian, a calla lily, the morning star. I do not deny that fractal objects can have considerable aesthetic value, only that fractal structure constitutes the very standard of merit. Whatever is behind the need to find a mathematical or scientific criterion of aesthetic merit? "It is proportion and symmetry, the golden ratio!" "No! It is constrained disorder with a fractal dimension of 1.3!" How easy it is to play the game of counterexample with such proclamations. Perhaps the urge to ground aesthetic judgment on scientific or mathematical facts stems from an anxiety that art would otherwise be a realm of merely subjective taste. But if we recognize that aesthetic value can be the subject of rational discussion—neither insulated from the facts that science describes nor collapsible into them—we may at last be able to become pluralists about beauty.

Fractals and Informational Complexity

Hawkins's fractal aesthetic also relies on a notion that fractal structures produce a level of complexity she sees as intrinsically valuable. For example, she bases her notion that chaos is beautiful on the fact that the Mandelbrot set generates astonishingly complex computer graphics (1995, 81). We have already seen in our discussion of Argyros just how problematic it is to assume that more complex works of art are superior, but Hawkins faces a slightly different problem. Early in her discussion, she invokes a definition of complexity that "involves our apprehension of the pleasures and difficulties involved in works manifesting a comparable richness of informational content and detail on all scales" to distinguish between classics and their spin-offs (13). For example, she asserts that publishers of romance novels "can send their authors a brief set of instructions from which they can construct successive novels. By comparison, a set of instructions informing authors how to construct a work on the order of *Paradise Lost* or *The Tempest* would be virtually interminable, requiring volumes far, far longer than the word-count of either work" (13).

Hawkins takes richness of informational content on all scales as an "accurate description of complexity in literature generally" (56). Although a popularized version of a classic work might follow the same general plot lines, the difference is the "greater and lesser developments of informational content and complexity of detail on every scale." So *Jurassic Park* the novel is less complex than *Paradise Lost*, and *Jurassic Park* the movie is less complex still. Hawkins gives an admirable reading of Milton's masterwork, but this does not prove that one could not give an equally virtuosic reading of the complexities of Crichton's novel. What about its place in the tradition of didactic literature that includes *Uncle Tom's Cabin*? What about its resonances with Arthur Conan Doyle's *The Lost World*? Why no mention of the recurring yet unpredictable pattern of calm punctuated by unpredictable dinosaur attacks? Far be it from me to argue that Crichton is every bit as much a literary genius as Milton. But can we really locate the difference in some measurement of complexity?

Here Hawkins, like many others who speak of chaotic systems as "rich sources of information," seems to be drawing indirectly on early work by chaos theory pioneer Robert Shaw. In an article published in 1981, Shaw discussed how the sensitive dependence at work in chaotic systems can serve to take random thermal fluctuations at the molecular level and translate them up to the macroscopic level, where we can observe them.[11] Because sensitive dependence exponentially magnifies any small vagueness in a system's initial state, a chaotic system can effectively act as a source of random numbers. And the mathematics of information theory tells us that nothing is so rich in information as random numbers. This follows from what is known as algorithmic complexity theory, which identifies the complexity of a string of numbers with the length of the instructions needed to generate that string (Chaitin 1977). A simple repeating series such as 121212...can be easily reproduced with a short instruction set: keep writing "12." A string of random digits, by contrast, cannot be effectively summarized; the shortest way to tell someone how to reproduce it would be simply to tell them to write it down symbol by symbol.

Literary information is not a formal property residing in the text alone, however; it also depends on the community's practices of interpretation (Fish 1995, 13) as well as the richness of its connections to its context. As Livingston puts it, "the 'best' literary text, by canonical or disciplinary standards, may well be the one that is most intertextually implicated—*thickest*—with other literary texts that come before, around, and after it" (1997, 4). Because of these intertextual resonances, simple works can be incredibly rich. Consider a haiku that can be transmitted with little bandwidth but conveys a wealth of meaning. And what of a novel such as *In the Labyrinth*, wherein an identical passage is repeated a number of times to different effect? Hawkins suffers from conflating the mathematical notion of "information content" with the "meaningful information" of a literary work.[12] The amount of meaning in a work of literature cannot be measured by the length of the instruction set needed to construct another message of comparable merit.

I draw on the work of Ira Livingston to conclude this brief against the fractal aesthetic with one final criticism: fractals do not provide images of timeless natural beauty but instead visualizations of the way contemporary sociopolitical power is coordinated across all scales of late capitalist society. Livingston maintains that "looking at how scientists are 'constructing' nature (and mathematics) in the image of postmodern power may be a necessary—politicizing—step in correcting the aestheticizing and scientizing that are rampant in the discourse of fractals and chaos" (1997, ix).[13] Drawing on the work of Michel Foucault (1979), who chillingly described the workings of institutional authority and control in terms of "discipline," Livingston characterizes modern and postmodern power as a "metastasizing sprawl of disciplinary networks." In this scenario, "every network is already a network of networks... by achieving a certain density of saturation it reveals each of its elements as networks as well. Disciplinarity is a plaid, a pattern of patterns that works not by being radiated from a center but by generating correspondences among nodes in multiple networks" (Livingston 1997, 22). If this chilling description of the way sociopolitical control penetrates every level of our lives rings even a little true, it surely throws some cold water on our enthusiasm for fractal images. And nothing can make them seem so unfashionable as calling them plaid.

One Last Bridge: Mimesis

Some literary scholars seek to locate a standard for evaluation in the formal properties of a work, such as its structure or complexity. Others make the aesthetic response of the audience into the touchstone for evaluation. We have seen both of these strategies at work in the borrowings from chaos theory employed by Argyros, Hawkins, and others. Some other standards do not seem to play much of a role in discussions of chaos, for example, the standard that locates artistic merit in the distinctive intent of the author. And many reject the entire enterprise of evaluating artistic merit as irrelevant, impossible, or hopelessly context bound. This rejectionism is closely associated with the loose movement known as postmodernism and is treated in the next chapter. But chaos seems especially well suited to one final realm of artistic judgment: the degree to which a work accurately represents the world. An appeal to this standard, the standard of mimesis, underlies some recent scholarship on chaos theory and literature.

Having found ubiquitous patterns of order and disorder in literature, Hawkins asks why there should be so much chaos in art "unless art reflects cognate mysteries in life?" (1995, 47). And having put forward a fractal aesthetic, she suggests that the concepts and images of chaos may be so widely appealing because they are good representations of the real world (167). "Mimesis" has been used to name this tendency to speak of art as reflecting and representing nature. Other scholars rely on the mimetic function of art to establish not only the relevance of chaos theory but the value of literature itself. For example, Mackey repeatedly returns to the possibility that the connections between chaos theory and *Ulysses* may show that Joyce has something important to say about real life (1999, 60). Demastes asserts that American theater has begun to reveal the nature of the world as dynamic and open-ended (1998, 117). According to him, this turn toward "the dynamic, nonlinear, and evolving chaos of life" may give us "a more accurate depiction of nature and a clearer vision of how to function within that universe" (128). And Larry Short points out that the quality Kant identified as "purposiveness" is related to naturalness, and naturalness is captured to some extent by fractal mathematics. So fractal structures can contribute to the experience of beauty and provide an "objective referent" for it, though not a sufficient condition (1991, 352). For these writers, the venerable and contested concept of mimesis serves as a resource for building argumentative bridges from facts to values.

Attempts to read social implications off nature often end up simply reading social preconceptions and preferences into nature. But the solution is not to condemn this interpenetration and seek quarantine, but rather to foster more critical and self-aware interaction. Borrowing from the natural sciences can help articulate a value and argue for why it is valuable, but it is much more difficult to establish one criterion as the exclusive value for a particular field. Chaos theory may help make sense of how adaptability is desirable in certain situations, or how complex fractal structure contributes to the beauty of certain works of art, but it cannot foreclose all evaluative discussion by enshrining flexibility or complexity as singular, sufficient, and paramount virtues. Again we find the need for ground between isolation and collapse—the natural sciences can and should play a substantive role in evaluative inquiry, but they should not be used to end such inquiry with definitive pronouncements.

9   Postmodern Chaos and the Challenge of Pluralism

Chaos theory and postmodernism both hit the academic big time in the 1980s and 1990s. Jean-François Lyotard's highly influential *The Postmodern Condition: A Report on Knowledge* was translated into English in 1984, and science journalist James Gleick's best-selling *Chaos: Making a New Science* (1987) brought nonlinear dynamics to a wide audience in an accessible, virtually math-free format. Lyotard's announcement of the arrival of a "postmodern science" of instability, nonlinearity, and fractals helped to link these two fashionable multidisciplinary frameworks almost from their first appearance. This connection between postmodernism and chaos theory was asserted in the catalog of a 1989 art exhibit at the New Museum of Contemporary Art: "With its reliance on computer simulation, its themes of discontinuity, unpredictability, fragmentation, and chance, chaos science at once participates in, and provides models for describing, a more widespread shift in the realm of culture. Like postmodernism in the arts and poststructuralism in the humanities, chaos science traces its lineage to the sixties" (New Museum 1989, n.p.). And the editor of a volume on chaos and criminology claims that "chaos theory has emerged as one of the key threads of postmodernist analysis" and that it is "one of the constitutive threads of postmodernist thought" (Milovanovic 1997a, vii; 1997b, 195; see also Williams and Arrigo 2002, 23).

In this chapter I survey some of the supposed connections between postmodernism and chaos theory in literature, literary theory, and the law. The examination of these purported connections provides an opportunity to revisit many of the themes from earlier chapters: the rhetorical power of disciplinary prestige, the many uses of metaphorical borrowing, and the role of science in evaluative debates. And because the connection between chaos theory and postmodernism played a significant role in the so-called science wars of the 1990s, an investigation of this connection provides a useful occasion for reflecting on the relationships between disciplines. I begin this chapter by looking at some examples of analogies and mismatches between chaos and postmodern literature, and then consider claims of resonances and conflicts between chaos theory and the movement within postmodern literary theory known as deconstruction. I then move to an examination of the role of chaos theory in postmodern legal thought, especially the work known as critical legal studies. The scrutiny of borrowed knowledge—in this and previous chapters—invites a reassessment of the relationships between the various knowledge-making disciplines, and I conclude by arguing once more for disciplinary pluralism. A pluralistic openness to a variety of approaches is helpful for more than the specific task of understanding borrowings from chaos theory; it can reshape our conception of knowledge making in general.

For the purposes of this chapter, the term "postmodernism" names a cluster of attitudes and approaches marked by a detachment (whether playful or despairing) from certain conceptions central to modernity. Among the notions that postmodernists question, reject, or treat ironically are progress, consensus, the unified subject, value-free rationality, and utopian freedom. As a roughly defined cluster of concepts and postures, postmodernism has shown up throughout contemporary culture, from architecture to philosophy. But for the purpose of this chapter I begin with definitions given by two of the most important figures articulating a connection between chaos theory and postmodernism: Lyotard and Hayles. Lyotard characterizes postmodernism largely in terms of the loss (or exhaustion) of grand narratives—the stories we tell that legitimize knowledge and society (1984). For Hayles, postmodernism resides in the awareness that supposedly essential components of human experience are contingent and constructed (1990). Of course, these definitions of postmodernism have never become definitive, and I certainly do not attempt to resolve the disputes about what postmodernism is (or was). But Lyotard and Hayles provide influential first approximations, and diverging views are noted as I proceed.

With these working definitions of postmodernism in hand, we can survey a wide variety of relationships that have been claimed to hold between this cultural formation and the science of chaos theory. Joseph Conte holds that early postmodern writings in literature and theory anticipated the findings of nonlinear dynamics, and William Demastes claims that the recent science confirms what artists and critics have been saying for years. On the opposing side, Alexander Argyros argues that chaos theory actually disproves central tenets of postmodern thought. Some authors are content simply to note similarities between certain elements of postmodern culture and chaos theory, whereas others such as Matheson and Kirchner take pains to show that these similarities are exaggerated or illusory. On the basis of the suggested similarities, Hayles proposes that the products of science and of art respond to a common underlying cultural situation, whereas others question the notion of common cultural causation. Let us turn first to the field of postmodern literature and see how these varied claims play out.

Postmodernism in Literature

Resonances and Mismatches in Conte

Joseph Conte's *Design and Debris: A Chaotics of Postmodern American Fiction* makes a sustained argument for commonalities between some contemporary literature and aspects of chaos theory. Conte describes this relationship in a wide variety of ways, beginning with the modest claim that recent novels "anticipate developments in the science of orderly disorder" (2002, ix). According to Conte, postmodern authors destabilize order and revel in unpredictability, displaying an "affinity" for disorder (8). Because of this affinity, "the postmodern writer both anticipates and fervently illustrates theories of chaotic behavior and unpredictability before their adoption by the academy as a legitimized narrative" (10). Moving beyond anticipation, he states that the two fields display "coeval development" (2) or "correlation" (3), and he goes on to characterize this correlation as not merely an intriguing accident but as a "homologous development" or "convergence of inquiry" that qualifies as "inevitable" (3). Later, Conte suggests that "the world-making of postmodern fiction sometimes anticipates and sometimes confirms the scientific theories of chaos and complexity" (27). Surely the notion that works of fiction can "confirm" scientific theories represents a bold claim, and it is important to examine the evidence that Conte offers in order to determine the import of the affiliations he presents.

Where do these supposed commonalities reside? Conte locates specific metaphorical resonances between strange attractors and elements in the work of the postmodern authors he identifies as "proceduralists." This group includes writers such as Barth, Calvino, Perec, and Sorrentino, who "formulate a plan comprised of arbitrary and exacting rules, carrying it out in spite of—or in anticipation of—the narrative consequences" (2002, 27). According to Conte's reading, novels such as Hawkes's *Travesty* (1976), Gilbert Sorrentino's *Pack of Lies* (1997), and Coover's *The Universal Baseball Association, Inc., J. Henry Waugh, Prop.* (1968) are works in which "an undisclosed pattern becomes apparent in disorder; they are novels that reveal an immanent design in the fractious conditions they describe. They are homologous in form to the deeply encoded structures that are the subject of nonlinear dynamics" (Conte 2002, 4). Thus, "postmodern proceduralism bears comparison to the strange-attractor branch of chaotics. Often simple, deterministic rules can produce complex results in a nonlinear system, as these narratives demonstrate" (28).

Lacking familiarity with all these works as well as experience in literary interpretation, I do not endeavor to evaluate the adequacy of Conte's suggested categorization of these authors or his characterization of their techniques. But we can still identify some specific places where his identification of similarities proves convincing. For example, just as we saw some cases of metaphorical sensitive dependence in literary works in chapter 6, Conte notes that the narrator in Sorrentino's *Rose Theater* claims that "rigorous attention to the most pedestrian details of human relationships may yield surprising data if not any decent 'yarns'.... Some may discern in it, perhaps, the presence of what Hans Dietrich Stöffel in his *De praestigiis amoris* (Brussels, 1884) defined as '[a] large mess (*perturbatio*) developing as if self-generated out of [a] small one'" (Conte 2002, 94, quoting Sorrentino 1987, 207). And the character Dr. Peloris in Coover's *Pricksongs and Descants* hearkens to the appeal of strange attractors when he declares that "even nonpattern eventually betrays a secret system" (Conte 2002, 141).

Unfortunately, Conte sometimes conflates chaos theory with other scientific theories such as quantum mechanics that also pose challenges to predictability. For example, he states that "the turn toward unpredictability in postmodern science establishes an affinity with narrative that classical science, in its demand for deterministic results, could not permit" (2002, 3). But this assertion ignores the fact that chaos theory, unlike quantum theory as generally interpreted, is fully deterministic. It is hard to read quantum theory as postmodern when it was developed in the late 1920s and admired by the very pre-postmodern logical positivists. Similarly, Conte misidentifies elements of chaos theory in Kathy Acker's *The Empire of the Senseless*. When a character in Acker's novel says, "If there is any variability to reality—functions which cannot be both exactly and simultaneously measured—reality must simultaneously be ordered and chaotic or simultaneously knowable and unknowable by humans" (1988, 102–3, as quoted in Conte 2002, 64), this statement seems to refer to quantum mechanical uncertainty relations and the fact that uncertainty means we can only know probabilities.[1]

Conte's detailed readings of specific works also suffer from some serious metaphorical mismatches. In his extended treatment of Hawkes's novel *Travesty*, for instance, Conte finds the phrase "the utter harmony between design and debris" to be an apt expression of the insights of chaos theory, but he faces a problem of fit (2002, 37, quoting Hawkes 1976, 17). For Conte, this phrase suggests that "if design inevitably surrenders to debris, debris inevitably reveals its innate design" (Hawkes 1976, 59), but chaos theory posits no such inevitability. In Hawkes's novel, a jealous husband speeds toward an elaborately planned collision, and his passenger is the man with whom he believes his wife has betrayed him. The main character appears fixated on images of a sudden crash and explosion, with parts scattered at random, but this is not at all an image of low-dimensional deterministic chaos or "the instantaneous phase transition from laminar to turbulent flow" (Conte 2002, 36). Explosions are not bounded systems; they make predictions impossible by sending nearby points hurtling far away, not by stretching and folding nearby trajectories into intricate patterns. So Conte errs when he casts the protagonist's imaginings of a hidden plan underlying the scattered fragments of the car wreck as Hawkes's demonstration "that the postmodern novelist apprehends one of the central tenets of chaos theory not as a figure of speech but as a physical principle" (39). Finally, Hawkes's protagonist rejects predictability because, on Conte's reading, "the variables are too many for such sureties" (43), but the proliferation of variables has nothing to do with the source of unpredictability in low-dimensional chaotic systems. Some of the apparent parallels between this novel and chaos theory ultimately rest on mistaken readings of the science.

Similar problems plague Conte's treatment of Coover's *Universal Baseball Association* novel. The complexity of the tabletop baseball league developed by the protagonist arises from the calculated combination of chance and design: dice are rolled to generate a probabilistic spread, and tables dictate the outcome results in the game world. This method is not analogous to chaos theory, but rather to the way thermodynamics and quantum mechanics recover large-scale patterns from intrinsically random systems. When the randomness of dice counterbalances the controlling rule-charts, we do not have randomness arising from deterministic rules themselves. Although it is possible to do a close reading of a novel by using the terminology of chaos theory, that reading fails to be convincing when the chosen source of metaphorical insight proves less apt than an older scientific theory. Many of Conte's readings bring out valuable elements of the texts, but his claims of ineluctable convergence or confirmation are seriously undercut by problems of weakness of fit.[2]

Dichotomy Busting in Demastes and Pynchon

Although my examination of Conte's work focuses on issues of fit, William Demastes's book *Theatre of Chaos: Beyond Absurdism, into Orderly Disorder* illustrates a broader range of the issues that arise when borrowing from chaos theory to examine postmodern literary works. His work invokes the disciplinary prestige of the natural sciences not only to understand contemporary plays and clarify their import but to provide "validation" and "demarginalization" for their ideas (1998, xiii). Although recognizing that many of these plays present ideas about order and disorder that may not "sound terribly new," he believes that the rhetorical resources provided by a new science can explain these notions "with a language and a system appropriate to postmodern experience" (143). In this way he signals a recognition of the persuasive power of the new discussed in chapter 3.

Demastes presents a complicated account of the interactions between science and the rest of culture which makes room for the claims that art anticipates science and science validates art. Metaphorical defamiliarization plays a crucial role in his account, for Demastes says that the cultural awakenings of artists can help "soften the soil of a culture encrusted by a spirit of neo-Newtonian reason," to allow the new scientific vision of nonlinear dynamics, which then in turn shows that the visionaries were right (1998, 77).[3] But Demastes finds chaos theory useful not only for defamiliarizing traditional conceptions of order; it also restructures the target field by breaking open a dichotomy whereby playwrights have left us "to choose between existence represented as strict linear determinism or utter randomness" (104). According to his (admittedly simplified) account of recent theater history, "dramatic naturalism of the late nineteenth century created a theatrical dynasty of technique and

philosophy based on a strict Newtonian causal order" (xv).[4] In opposition to the dominance of this approach, works by those identified with
the theater of the absurd posited a universe of "nonrational, incoherent, and incomprehensible randomness" (56) as well as "stagnation"
and "entropic doom" (104).

In the face of this dichotomy, Demastes posits the recent emergence
of "a hybrid form, naturalist/realist in appearance but absurdist in sym-
pathies" (105). He finds an important precursor to these works in those
of Samuel Beckett, who—unlike other absurdists—strives to create a
hybrid that integrates order and chaos (58). Similarly in more recent
postmodern works, "keystone aspects of both naturalism and absurdism coexist," but these plays ultimately "argue for a dynamic interplay
between order and disorder rather than, say, either an alternating emphasis" (107). Postmodern writers such as Shepard and Rabe reject the
strictures of naturalistic narrative lines, but they also see the absurdist's
rejection of "language, logic, and form" as "merely counterproductive
radicalism" (120). For example, in the afterword to his play *Hurlyburly*
(1985, 162), David Rabe speaks of "the 'realistic' or 'well-made' play
[as] . . . that form which thinks that cause and effect are proportionate
and clearly apparent . . . that one thing leads to another in a rational, mechanical way, a kind of Newtonian clock of a play . . . the substitution of
the devices of logic for the powerful sweeps of pattern and energy that is
our lives" (as quoted in Demastes 1998, 108). Here we clearly see that
the image of patterned disorder borrowed from chaos theory can be useful for Demastes in articulating an antidote against an image of dramatic
order based upon an earlier borrowing from physics. Indeed, Rabe's play
includes the phrase "the pattern in the randomness" (Rabe 1985, 89, as
quoted in Demastes 1998, 116).[5]

If we grant him the broad sweep of his history of twentieth-century
theater, we can accept that Demastes has borrowed from chaos theory to
construct a useful antidote for the opposition between naturalistic order
and absurdist disorder. However, we must acknowledge that there is
more than one way to bust a dichotomy. Demastes himself characterizes
his challenge to the rigid split between order and chaos in a number of
ways, and not all of these provide a coherent match with the structure
provided by chaos theory. For instance, at some points Demastes places
order and disorder on a continuum, which surely challenges the notion
that nothing can be both somewhat ordered and somewhat disordered
(1998, xi). But chaos theory seems to present us not with a continuum
but instead with a hybrid, because systems characterized by strange
attractors are at the same time both fully deterministic and effective

sources of randomness. A chaotic system does not just *usually* obey causal laws; neither is it only *sometimes* unpredictable. Nor does chaos theory reveal that order and disorder interact and alternate, with each one arising from its opposite in turn (16, 63). Third terms, shades of gray, hybrids, self-contradictions—rhetorical scholars have cataloged any number of structures that can provide an alternative for a field set up with a strict either/or (see Fahnestock 1999, 89). Here one can apply the criteria of distinctiveness and productivity: does chaos theory offer resources for restructuring that are unavailable elsewhere? And to what extent can the structures of chaos theory be carried over to extend the metaphor beyond the initial breaking of the binary opposition? In the case of Demastes, the productivity of chaos as a metaphor runs out fairly quickly.

The work of Thomas Pynchon has also been held up as an example of how literature anticipated a nondichotomous relationship between order and disorder. Unfortunately, some readings of Pynchon suffer not only from a misunderstanding of nonlinear dynamics but also from lack of attention to the novels themselves. For example, attorney Jay Moran has characterized the novel *Gravity's Rainbow* as Pynchon's "most sophisticated attempt at using chaos theory to delineate the social landscape," even though it was published in 1973 (Moran 1997, 177). Many have seen Pynchon as posing a choice between the paranoia of total determinism and the antiparanoid dissolution of identity in utter randomness. Slethaug (2000) returns again and again to Thomas Pynchon's novella *The Crying of Lot 49* as a "pessimistic" meditation on the opposition between chaos and order, meaninglessness and paranoia. But Slethaug never acknowledges that entropy is not the same as deterministic chaos, or that chaos theory actually undermines the pessimistic dichotomy, or that Pynchon repudiates the dichotomy himself (where Oedipa muses that excluded middles are "bad shit") (Pynchon 1966, 181). In his attack on the role of postmodernism in legal scholarship, Moran also fails to note that Pynchon, or at least his protagonist in *Lot 49*, rejects the opposition of paranoia versus meaninglessness (1997, 186). Furthermore, he identifies the "chaos" in that novella with a senseless mass of details and "entropic forces" (179), missing the point that chaotic behavior needs to be distinguished from a mere thermodynamic mess.

Joseph Conte correctly notes that Pynchon presents an alternative besides the two forks of the dichotomy (2002, 178). But because both Pynchon and Conte speak of the logician's law of the "excluded middle," the match with chaos theory becomes less convincing. Aristotle's law excluded any middle term between true and false, mandating that every statement must be at least one or the other. But although this law forbids a statement that is *neither* true nor false, it does not prohibit a statement from being *both* true and false—that job falls to the law of noncontradiction. To the extent that chaotic systems are both deterministic and random, both ordered and disorderly, they metaphorically challenge noncontradiction rather than signaling the triumphant return of the excluded middle. Chaos theory proves useful indeed for challenging dichotomies, especially the dichotomy between order and randomness.[6] But it appears that sometimes the persuasive power of the new outweighs any distinctive conceptual work that chaos can do.[7]

Modernism or Postmodernism?

Postmodern literature cannot claim exclusive rights to affinities with chaos theory. Recall the resonances that Peter Mackey and others have found with such modernist works of literature as Joyce's *Ulysses*, not to mention Hawkins's discussions of Shakespeare and Milton. Furthermore, the novelist John Barth has constructed an extended argument that the romantic notion of the arabesque presages the chaotic elements in postmodern literature. By "arabesque" he means an artfully structured confusion—an orderly disorder—within a narrative frame marked by self-similarities at multiple-scale levels (Barth 1995, 344). Barth writes that the aesthetic of the romantic arabesque, which draws inspiration from the convolutions of Cervantes as well as the intricacies of so-called Persian carpets, "seems to me to be potentially revalidated—refreshed, reinforced, replenished—by contemporary chaos theory" (1995, 289).[8] So what are we to make of the fact that researchers have found similarities between chaos theory and postmodernism, given that similarities have also been found with virtually every literary tradition of the last few centuries? Does this show that English professors can get away with anything? Or does it show that modernism and postmodernism are not really so different, since both explore themes of complex meaning-making systems?

First, let us examine what, if anything, is the difference between modern and postmodern literature. According to Conte, the function of the modernist artist is "to impose order and coherence on a disorderly, random, and inchoate world," whereas postmodernists seek to destabilize order and revel in unpredictability (2002, 7–8). Yet some consider authors such as Coover, Hawkes, and Barth to be "late-modernists" rather than postmodernists because they retain a high regard for the structuring projects of modernism (Barth 1995, 295). This problem becomes incredibly tangled when we consider an essay by Hans Hahn (1954), a member of the Vienna Circle. In this essay on geometrical knowledge (based on an even earlier, unpublished lecture), Hahn used figures that now would be called fractals to argue against Kantian notions of mathematical intuition. If anyone is not postmodern, it would be a logical positivist. So does this essay demonstrate that fractal geometry is not especially consonant with postmodernism, because it was embraced by an arch-modernist philosopher? Or does it show that a piece of mathematics can have very different philosophical and cultural interpretative spins placed upon it?

Some opponents of postmodernism provide examples of this interpretive spin when they turn chaos theory into something quite surprising in their attempts to reassert a more traditional view of science. For instance, Matheson and Kirchhoff say that postmodernism "can be hastily summarized as a crisis of authority in our standards of knowledge and a denial of objective existence"; they then go on to assert that, in opposition to postmodernism, chaos theory gives us a modernist picture of determinate yet unknowable nature: a "clockwork and beautifully harmonious reality" (1997, 36). Yet, as described in the introduction, chaotic systems are precisely not clockwork. For Thomas Rice, the "restoration of objective knowledge and determinism" establishes the importance of chaos theory for our culture and for reading literature (1997, 107). Oddly enough, Rice says that chaos theory "reaffirms certainty in the realm of science" (85). Gordon Slethaug offers an insightful criticism of literary scholars such as Rice when he notes that the finding of patterns within apparent randomness can lessen anxiety, but "this coherence, pattern, or order implicit in, or deriving from, chaos may simply be the siren song of a new Platonism or a nineteenth-century Emersonian optimism, declaring that behind every infraction is a rule and that underlying uncertainty is certainty" (2000, xiii). Although champions of ideals such as certainty or objectivity find vindication in chaos theory, postmodernists skeptical of those ideals borrow from the same science to argue the opposite point.[9]

Perhaps the difference between modernism and postmodernism is not a difference in techniques of invention or composition, such as fractal structure, fragmentation, or wandering narrative lines. Neither can we locate it in differences of theme, such as the interplay of order and disorder or attention to the trivia of everyday life. Instead, we should situate the distinction in terms of a different attitude toward the goals and pretensions of the Enlightenment (see Hayles 1990, 295). If postmodernists are less likely to take seriously the heroic projects of modernist literature, this posture does not mean that they have utterly escaped from modernity. Chaos theory is not a radical break with the tradition of Western science, just as romanticism and postmodernism are not radical breaks with the Enlightenment. So finding literary resonances with chaos theory does not necessarily establish a link to either modernism or postmodernism, and it's not particularly hard to do either, given the tremendous resonances of words like "chaos," as noted earlier. Livingston diagnoses the situation well when he says that "looking back to the literary anti-Newtonianism of the Romantics leads to a scenario whereby quantum physics (as a marker of modernity) and later chaos or complexity theory (as a postmodern marker) can then be taken as science's incorporation of (or liberation by) its disciplinary Others, at least insofar as these movements legitimize anti-Newtonianism from within science rather than without" (1997, 40; see also van Peer 1997, 48). As Barth admonishes, showing "affinities" is not enough; the important question is, What can we make of them? (1995, 290). Chaos theory can serve as a valuable resource for elucidating what is going on in a work of art, yet chaos theory itself has multiple meanings, sometimes encouraging us to discern hopeful signs of order and sometimes enabling us to appreciate the possibilities of randomness. We can find fractals, fragmentation, the play of chance, and the momentous consequences of minor events in Joyce as well as in Pynchon, but with different inflections and emphases: Are we looking for heroic action in the face of interpenetrating order and disorder, or ironic celebration of their mixed-up-ness? Although chaos theory cannot conclusively validate or disconfirm an aesthetic judgment, it can motivate the acceptance of new ways of looking at literature.

Chaos and Postmodern Literary Theory

Deconstruction

Turning now from postmodern literature to postmodern literary theory, I concentrate on one especially salient thread: the interpretive approach known as deconstruction. The relationship between chaos theory and deconstruction has been characterized as one of surprising similarity by Katherine Hayles in *Chaos Bound* (1990), whereas Alexander Argyros characterizes it as one of factual incompatibility in his book *A Blessed Rage for Order* (1991). In their book *Higher Superstition* (1994), Paul Gross and Norman Levitt assert that the two have nothing to do with each other. How can we determine the relationship between a scientific discipline and a literary-philosophical method? I begin by saying something about what deconstruction is.[10]

A deconstructionist might chuckle at any attempt to define deconstruction once and for all, but for my purposes here it is helpful to spell out some basics. Deconstruction began as an approach to reading the history of philosophy, developed by Jacques Derrida. It later became associated with an informal school of literary criticism at Yale University, after which it consolidated into a critical methodology. Since the 1990s, "to deconstruct" has become a (slightly dated) synonym for "to criticize."[11] As initially formulated by Derrida, deconstruction often worked by taking one of the fundamental oppositions of Western philosophy (reality versus appearance, reason versus emotion, etc.) and showing how one of the two terms was privileged, yet never completely free of its dependence on that which it sought to exclude. Numerous examples of this strategy show how the valorized term on the "inside" of a conceptual dichotomy secretly relies on its "outside" (or its "other"). For example, recall the discussion about the distinction between literal and figurative language in chapter 5. Those who insist on clear, literal language run up against the fact that all language involves using some things (sounds, marks, etc.) to stand for other, very different things (trees, rocks, concepts, etc.). So even the simplest literal declaration relies on something like metaphor (Derrida 1982). Derrida's deconstructionist readings show again and again how great philosophers have ended up saying just what they did not intend, because meaning is never fully determinate and no term can ever be nailed down to its referent in a way that guarantees fixed, transparent reference.

Hayles

N. Katherine Hayles, a professor of English at UCLA who also holds a master's degree in chemistry, published her book *Chaos Bound: Orderly Disorder in Contemporary Literature and Science* in 1990—only three years after Gleick's best seller. Much of this text is not about chaos theory narrowly construed as the study of low-dimensional deterministic chaos and strange attractors, for it also deals with Prigogine and with conceptions of disorder associated with thermodynamic randomness and information theory. Hayles engages in broad, speculative, and occasionally highly insightful cultural criticism that "reads" science as well as literature. In the sections most relevant for our discussion here, Hayles postulates that both chaos theory and deconstruction arose within the same matrix of social and economic influences and constraints—the same "ecology of ideas."[12] In this way she seeks to provide part of an answer to the question of why chaos theory blossomed into widespread popularity when it did. I set aside this question, which would require a detailed historical and sociological investigation, and instead focus on the many ways that Hayles's project can illuminate issues involved in cross-disciplinary borrowing.

Unlike many who borrow from chaos theory to argue for or against some interpretive stance or theoretical position, Hayles uses both chaos theory and deconstruction to make a larger point about contemporary culture in general. Thomas Rice claims that Hayles sees chaos theory as an "affirmation" of "her poststructuralist reading of literature" (Rice 1997, 154), but Hayles is not primarily borrowing from the natural sciences to support her literary interpretations. Instead, she treats the natural sciences as a source of data for her broader observations. Nonetheless, her overall argument rests in part on the assertion of a metaphorical convergence between chaos theory and deconstruction, and this claimed similarity has proven influential for later work at the crossroads of science and literature.[13] Despite the fact that some have simply dismissed her explanation in terms of a common cultural formation as "a bizarre thesis," (Gross and Levitt 1994, 99), it is worthwhile to examine this purported cross-disciplinary resonance in some detail.

Resonances. Hayles argues that chaos theory and deconstruction share "assumptions and methodologies," and that these similarities "can hardly be explained without the assumption that both are part of a common episteme" (1990, 176). According to her construal of deconstruction, "the permeation of any text by an indefinite and potentially infinite number of texts implies that meaning is always already indeterminate. Because all texts are necessarily constructed through iteration (that is, through the incremental repetition of words in slightly displaced contexts), indeterminacy inheres in writing's very essence" (181). Two crucial terms in this passage—"indeterminacy" and "iteration"—are called upon to establish a connection to chaos theory. Indeterminacy serves as the shared assumption, and iteration serves as the shared methodology mentioned previously.[14]

When we examine these features in detail, the metaphorical character of the comparison becomes accessible to criticism. Looking first at the methodology designated "iteration," Hayles claims that Derrida's practice of "repeating Rousseau's language with incremental differences becomes a way to unfold and make visible the inherent contradictions upon which the text's dialectic is based. This iterative procedure produces the undecidables that radically destabilize meaning. . . . The goal of iteration is thus to make visible the lack of ground for the alleged originary difference, thus rendering all subsequent distinctions indeterminate"

(1990, 182). Hayles claims that this reconstruction of deconstructive methodology is "strikingly similar to the mathematical techniques of chaos theory" (183). Note that I am leaving aside the highly problematic (and trivially self-defeating) question of whether Hayles is correct in her attempt to extract the essential methodology of deconstruction.

The claimed similarity between Derridean deconstruction and chaos theory is based on three parallels: "they agree that bounded, deterministic systems can nevertheless be chaotic; they both employ iteration and emphasize folds; and they concur that originary or initial conditions cannot be specified exactly" (Hayles 1990, 184). These suggested parallels rely on a nonliteral, or broadly metaphorical, appropriation of nonlinear dynamics. The systems studied by deconstruction are not deterministic systems because they are not governed by strict and unchanging mathematical equations; the objects of literary analysis are not dynamical systems amenable to quantitative techniques. Indeed, the fact that a phrase can be exactly repeated in a text with a different effect, as in the work of Robbe-Grillet, could be taken as evidence that such systems are not deterministic at all. The iteration at work in a text, or in textual criticism, does not apply an identical procedure to the output of each previous application of the procedure. The initial conditions for a dynamical system do not serve any originary meaning-bestowing function. But pointing out the nonliteral use of the terms "deterministic system," "iteration," and "initial conditions" is not in itself a criticism of Hayles's claims for a similarity between chaos theory and deconstruction. Indeed, there is a traditional sense of literary texts as well-ordered, regular systems of meaning—almost clocklike, if not entirely "Newtonian." So Hayles is responding in part to a heritage of earlier metaphorical borrowings.[15]

Against Resonances. While claiming a strong similarity, Hayles also acknowledges substantial differences between deconstruction and chaos theory. For one thing, deconstruction seems to treat all texts as chaotic, but nonlinear dynamics recognizes the occurrence of predictable, nonchaotic systems. For another, chaos theory can hardly be portrayed as antiscientific, and science looks for order whereas deconstruction undercuts it; deconstruction celebrates disorder and the escape from mastery, but science does not (Hayles 1990, 292; see also Livingston 1997, 40). Hayles recognizes that chaos theory extends the reach of scientific, even Newtonian, analysis, and "in this respect it is profoundly unlike most poststructuralist literary theories, especially deconstruction" (1990, 15). And she expresses a clear mistrust of Lyotard's claim that "chaos theory provides confirmation from within the physical sciences that totalizing perspectives are no longer valid" (15).[16] Unfortunately, Hayles does not often receive credit for explicitly recognizing these disanalogies. Somehow, her claim of similarity between chaos theory and deconstruction is "bizarre" enough not to merit evenhanded treatment.

And evenhandedness demands that I now turn to a critique of her metaphorical treatment of chaos theory. Hayles seeks to show a similarity in terms of chaotic behavior, iteration, and the inaccessibility of initial conditions, with an eye toward demonstrating a broad cultural parallel rather than arguing for a methodological revision. She uses chaos theory as a kind of cultural data to show similar responses to common constraints, rather than as argumentative evidence to show the need for a new interpretive method. In speaking of "cultural data" I mean that Hayles appeals to the phenomenon of "nonlinear dynamics," the research program, rather than appealing to the phenomena studied by that research program. But the evidence offered to support the thesis of a common "ecology of ideas" rests on metaphorical similarities with too much flexibility. Hayles interprets terms such as "deterministic systems," "iteration," and "initial conditions" in such a way that nonchaotic systems have every bit as much similarity to deconstruction as chaotic ones.

The first parallel Hayles suggests is that both Derrida and chaos theory challenge the binary opposition between order and disorder, agreeing that "bounded, deterministic systems can nevertheless be chaotic" (1990, 184; see also 17). Hayles points out that the methodology of deconstruction yields undecidables in a *regulated* way, "in the sense that its production of undecidables is not a capricious exercise but a rigorous exposition of the text's inherent indeterminacies" (183; see also Armour 1999, 77). Derrida speaks of "the system of a writing and of a reading which we know is ordered around its own blind spot," and his reading of Rousseau shows how such small indeterminacies generate wild, uncontrollable swings of meaning (Derrida 1976, 164). Certainly one of the most striking results from chaos theory is that strictly deterministic systems can exhibit complex and unpredictable behavior, and its usefulness in breaking down the dichotomy between order and disorder has already been noted in a number of places. But the criterion of distinctiveness proves important here: chaos theory is not the only, nor the first, example of a science that brings randomness and order into a paradoxical juxtaposition. The science of thermodynamics followed a well-regulated path to the study of random and unpredictable behavior. Quantum mechanics undertakes a rigorous exposition of the inherent indeterminacies of subatomic particles. And Gödel provided an exemplary instance of a strictly defined system that yields undecidables. The parallel between chaos theory and deconstruction becomes much less striking when we notice similar parallels in research from earlier cultural moments. There is little need to hypothesize a distinctively postmodern ecology of ideas or shared set of constraints in the postwar era.

A similar problem arises for the second suggested parallel, which concerns the use of iteration. Although it is true that many scientists have studied chaos in iterated functions, the role of iteration becomes much less suggestive when we consider that almost all systems of differential equations can be studied with iterative techniques. Many computer programs for generating the trajectories of dynamical systems use time-step methods for approximating a solution to the system. This technique takes the state of the system at one moment, computes an approximate integration of the system for a small time-step, and then repeats the procedure over again using the new final state as the next initial state. But there is nothing especially chaotic about this methodology—it had been used well before the development of chaos theory. Thus, although iteration characterizes a method used to numerically integrate a set of differential equations describing a chaotic system in order to study it, there is nothing especially iterative about chaotic behavior itself.[17]

Instead of positing a similarity with deconstructive method, Hayles's argument regarding iteration would find a better parallel between chaotic dynamics and Derrida's views on the structure of metaphor and signification. Hayles refers to Derrida's *Of Grammatology* for a discussion of iteration, but what we find on those pages (1976, 157–62) is a treatment of the problem of the method of reading, together with the now-infamous statement, "There is nothing outside the text." We also find a discussion of a chain of substitutable themes, including the "supplement," and an "indefinitely multiplied structure," which leads us to gaze into the abyss when we seek to follow signifiers back to an original extratextual signified (163). The structure Derrida describes seems to be that of infinite regress, wherein signifiers continue reaching back to some further repetition of metaphor or some further attempt to arrive at a transcendental signified. In contrast, the iterative feedback used to study some chaotic systems reaches forward in time, repeatedly reapplying a formula to the outcome of a previous application. This method of iteration defines the operation of difference equations, but it is in no way responsible for the chaos of some such equations.

Unlike iterations, the notion of "folds" may present a more robust similarity with chaos theory. As Hayles states, the "trace" discussed by

Derrida is like a fold—it creates and effaces an origin. And Kenneth Knoespel has observed that "deconstruction plays with the possibility that the inscriptions of ordinary language work as a kind of phase space being stretched and folded on the rectangle of the written or printed page" (1991, 108). The operation of stretching and folding is essential to chaotic dynamics, and the "fold" in the logistic map studied by Feigenbaum is responsible for the loss of knowledge about initial conditions. Yet Matheson and Kirchhoff peremptorily declare that "there is no plausible way in which either of these chaotic uses of 'fold' are remotely connected to the Derridean use. Concealment and topological folding have nothing important in common; this appears to be simply a case of a single word being used in two radically different ways" (1997, 34). Here we find a situation in which metaphorical uses are dismissed instead of being subjected to substantive criticism.

Talk of origins leads us to the third suggested parallel, the loss of information about initial conditions. In contrast to what Hayles claims, our lack of knowledge about origins is not the essential source of unpredictability in chaotic systems. For example, even with perfect knowledge of the initial conditions of a chaotic system, our inability accurately to compute and store information about its evolution renders long-term predictions impossible. The attempt to locate the essence of chaos in the unknowable origin misses the crucial role played by the dynamics of the system.[18]

Although it is true that researchers in chaos theory "concur" that "initial conditions cannot be specified exactly," this inability is not a result of chaotic dynamics. Limitations on the precision of measurement, whether due to the clumsiness of our instrumentation or the inherent indeterminacies of quantum systems, had been accepted by physicists for decades before chaos theory. Furthermore, in *any* system with friction, the passage of time results in the loss of information about initial conditions. Chaotic dynamics make these limitations matter in a way that many scientists may have liked to ignore, but there is no sense in which chaos is responsible for our lack of knowledge about initial conditions.[19] The flexibility with which Hayles uses terms from chaos theory sets a low threshold for establishing similarity with other fields, causing her claim that chaos theory and deconstruction share assumptions and methodologies to lose some of its distinctiveness (although it did not impede the persuasive effectiveness of her highly influential book). If the nonchaotic systems investigated in nineteenth- and early twentieth-century physics show the same signs of rigorous disorder, iterated differences, and loss of information about initial conditions, then the similarity between chaos theory and deconstruction can be explained without the assumption of a common episteme.

In their dismissal of Hayles, Gross and Levitt say that "Hayles' arguments, such as they are, are based on subjective and shoddy analogies, leaky metaphors, and (not unusual for work immersed in postmodern theory) flat and unsupported assertions" (1994, 99). But I have shown that, in addition to listing her mistakes, it is possible to criticize her arguments in detail, revealing both strengths and weaknesses. In response to these weaknesses, Matheson and Kirchhoff admit that one could argue that the similarity is suggestive though incomplete. "However," they declare, "any such suggestiveness dissolves upon examination. . . . We have shown that chaos and deconstruction are not isomorphic in any interesting sense" (1997, 35). Yet such sweeping negative appraisals are notoriously difficult to establish.

At this juncture, one might point out that Hayles offers a number of other pieces of evidence for correspondences and congruencies between postmodern theory and nonlinear dynamics. At the very least, the rapidly proliferating uses of the word "chaos" call to mind Derrida's discussion of dissemination. Furthermore, I have not dealt with Hayles's treatment of Barthes, Serres, and Foucault on the one hand, and Prigogine and Shannon on the other. My purpose has not been to rule on the general validity of Hayles's broadest claims but to examine the nature of the argumentation at work in her project—the treatment of broadly metaphorical resonances between work in the sciences and the humanities as data in need of a general explanation at the level of culture as a whole. So I have been following Knoespel's suggestion that we can compare chaos theory to deconstruction without trying to construct a grand story of deep underlying forces (1991, 102). My piecemeal, case-by-case examination of metaphorical resonances reveals that Hayles's work is amenable to, and deserving of, critical evaluation.

Argyros

Alexander Argyros argues for a very different kind of relationship between chaos theory and deconstruction, claiming that—setting aside some superficial similarities—he can demonstrate "the incompatibility of much contemporary research in science and mathematics with deconstructive principles" (1991, 34). Those who believe disciplines are isolated from each other might argue that the natural sciences can never prove or disprove a philosophical position. To the extent that Derrida is advancing a philosophy of language (and this of course is a highly problematic characterization of his project), it would be difficult for scientific results to be incompatible with his view. Difficult, but not impossible, according to the kind of naturalistic view described in chapter 2. After all, a philosophical position can derive plausibility from scientific results, and some general findings in physics have important metaphysical implications.[20] So Argyros's project may be unsuccessful, but it is not simply impossible. In this section, I examine three main arguments wherein Argyros uses chaos theory to combat deconstruction: he claims (1) that deconstruction maintains a dualism between order and disorder that chaos theory overcomes; (2) that deconstruction is antirealist whereas chaos theory reveals reality; and (3) that deconstruction is relativistic but chaos theory establishes universals. In the end, none of these arguments works, and I suggest that the problems stem not only from Argyros's misunderstandings of chaos theory (and deconstruction) but also from his failure to adopt a pluralist attitude toward the relationship between science and literary theory.

Deconstruction as Dualistic. No doubt should remain about whether chaos theory challenges the traditional opposition between order and randomness. If deconstruction truly upholds a rigid dichotomy between these terms, then it certainly would face a powerful challenge from the science of nonlinear dynamics. But given that deconstruction operates by seeking to subvert the dichotomies of Western metaphysics, for Argyros to claim that chaos theory undercuts an alleged "dualistic rigidity in Derrida's thought" (1991, 247) qualifies as audacious at best. How then does Argyros go about painting Derrida as a champion of the strict distinction between order and disorder? He relies heavily on the following quotation: "Language, however, is only one among those systems of *marks* that claim this curious tendency as their property: they *simultaneously* incline toward increasing the reserves of random indetermination *as well as* the capacity for coding and overcoding or, in other words, for control and self-regulation. Such competition between randomness and code disrupts the very systematicity of the system while it also, however, regulates the restless, unstable interplay of the system" (Smith and Kerrigan 1984, as quoted in Argyros 1991, 32, 246; emphasis in original). The second time he cites this passage, Argyros follows it by saying, "In many ways, Derrida is correct, of course. A certain play between chance and order is, in fact, crucial in most nonlinear systems" (246). But he contends that Derrida is wrong because "rather than undermining the systematicity of systems, the coordinated interplay between the erratic and the deterministic constitutes the systematicity of chaotic systems. Chaotic systems are not a field upon which the claims of order and

indeterminacy are disputed. In fact, it is precisely such a binary opposition, here between order and randomness, elsewhere between metaphysics and dissemination, that constitutes the most glaring dualistic
rigidity in Derrida's thought" (247).

Having suggested that Derrida holds a strict dichotomy between deterministic order and random disorder, Argyros makes the by-now-familiar
move of invoking chaos theory as presenting a third option. Curiously
enough, he states that chaotic behavior is "neither random nor deterministic" (1991, 254), but a few pages later he calls it "both random and
determined" (259). Either way, he makes it seem that "the simple opposition between order and randomness, which characterizes Derrida's
speculations on the subject, may be inadequate. Chaos may be a totally
new way of looking at evolving systems, one in which a kind of dynamic
stability is able to transcend the distinction between dissemination and
metaphysics" (245). But let us look more carefully at the extended quotation upon which Argyros relies. In the first place, notice that Derrida
is describing an *interplay* between disorder and information—hardly
evidence for any rigid dualism. Second, notice that the passage deals
with information rather than dynamics. And although information theory has been influential for some researchers in nonlinear dynamics, the
conflict between indeterminacy and control in information systems is not
reconciled in chaotic behavior. Argyros has conflated three oppositions:
indeterminacy and meaning, entropy and organization, and randomness
and determinism. Chaos theory overcomes the third of these oppositions, and self-organizing dissipative structures apparently challenge the
second. But Argyros has failed to establish a connection with the first
opposition—the very opposition Derrida treats.

Earlier in his book Argyros admits that deconstruction does not portray all meaning as random and indeterminate: "Derrida's purpose is
to argue for the structural necessity for chaos, not for its hegemony"
(1991, 32). He further admits that the deconstructionist "logic of the
mark" is "neither that of order nor randomness," and that Derrida "is
not interested in demonstrating that chaos came first, since that would
be as metaphysical a position as one which considered order to be original" (33).[21] Yet these admissions do not temper his later imputation of
dualism. Argyros says that deconstruction consistently highlights indeterminacy, because it has been traditionally devalued, and that Derrida's
prejudice for randomness is "something that can be inferred from his
choice of vocabulary and tone" (33–34). But Argyros exaggerates this
choice of emphasis so as to cast deconstruction as rigidly partisan. He admits that Derrida rejects strict binaries, he quotes Derrida talking about

pattern and disorder interpenetrating, but then he accuses Derrida of dichotomizing because he always takes the side of randomness.

As noted previously, Argyros locates another dualism in deconstruction such that "despite his insistence on the nonbinary nature of deconstructive concepts, Derrida's view of Being is essentially Manichean" (1991, 232). This good-versus-evil struggle can supposedly be found in the conflict between the traditional attempts of metaphysics to fix meaning and the fluid deconstructive workings of dissemination that disperse meanings in unpredictable ways. So Argyros once again invokes chaos theory against this supposed deconstructionist dualism: "If a stable attractor is, indeed, a mathematical-visual metaphor for metaphysics, then we should expect that Derridian textuality would be described by the absence of an attractor. . . . The phase space of dissemination would be fundamentally anarchic, periodically congealing into metaphysical stability but then quickly undermining its temporary hypostatization with unpredictable and discontinuous leaps of textual free play" (251). But deconstruction is not simply the opposite or the absence of metaphysics, just as postmodernism should never claim to have made a decisive break from modernism. The strange attractor, with its intricate mixture of order and randomness, might actually make a good metaphorical image for Derridean textuality.[22]

Where could Argyros have gotten the idea the Derrida simplemindedly privileges disorder over order? This simplistic reading of deconstruction becomes more plausible when we consider that a number of postmodern theorists have indeed fallen into a kind of dualistic thinking. Frankly, not all postmodernists are as good as Derrida at avoiding the trap whereby one simply inverts traditional oppositions by mechanically taking an excluded term and placing it at the center. Mathematician René Thom led the charge that postmodern scholars in the humanities celebrate disorder in his provocative piece titled "Stop Chance! Silence Noise!" (1983b), which was originally published in French in 1980. Thom's essay is a passionate denunciation of Ilya Prigogine and his admirers, who "outrageously glorify chance, noise, 'fluctuation.'" Their "fascination with randomness testifies to an antiscientific attitude par excellence. Moreover, in a large measure, it proceeds from a certain deliberate mental confusion, excusable in writers of literary formation" (11). Lest this seem like an intemperate exaggeration, note that Michel Serres, postmodern scholar of science and literature, has proclaimed, "Yes, disorder precedes order and only the first is real" (1974, 222, as quoted in Morin 1983, 23). And one of the most famous of the Yale deconstructionists, Paul de Man, has written: "nothing, whether deed, word, thought or text, ever happens in relation, positive or negative, to anything that precedes, follows or exists elsewhere, but only as a random event whose power . . . is due to the randomness of its occurrence" (1979, 69). Once it became a fashionable technique, a popularized version of deconstruction often became merely an exercise in "fragmenting signifiers for the sake of revealing the disorderliness and instability of signs and discourse in general" (Vanden Heuvel 1993, 259).

Yet, as we have seen earlier in the discussion of postmodern literature, a number of postmodernist scholars have perfectly well recognized the interpenetration of order and disorder in chaos theory (for other examples, see Morin 1983, 27; Vanden Heuvel 1993, 260). On the other hand, those postmodern theorists who insist on simply championing disorder while preserving a rigid opposition between order and randomness may indeed find that chaos theory provides a powerful antidote to their views. With respect to such dualism, Argyros actually makes good use of borrowed knowledge—he just aims his argument at the wrong target.

Deconstruction as Antirealist. The second main argument that Argyros offers to show that chaos theory is incompatible with deconstruction rests on his concern for scientific realism. Here, the term "realism" does not mean accurate literary representation or down-to-earth political commonsense but a metaphysical position that the natural sciences reveal the true nature of a physical world independent of human inquiry. Philosophers (and others) who abstain from this form of realism need not commit to a cartoonish subjectivism that says "Nature is whatever we say it is." One could hold, for instance, that the physical universe has no one true nature, or that its nature must remain unknowable. In any event, Argyros and several other critics of deconstruction view deconstruction as entailing the rejection of scientific realism. On this point they may well be correct, but it is quite another matter to show that chaos theory disproves deconstruction by establishing veridical contact with the nature of Reality. Let us examine the argument in more detail to see how Argyros's attempt to extract metaphysical conclusions directly from nonlinear dynamics overreaches.

Argyros portrays Derrida's famous statement that "there is nothing outside the text" as crucial evidence of his antipathy for scientific realism (1991, 21). According to Argyros, this aphorism establishes deconstruction as a kind of constructivism that views the entire world as simply the product of our consciousness—a clear denial that the natural sciences reveal the truth about an independent reality. To oppose such a view,

Argyros cites the familiar argument for realism that takes the form of an inference to the best explanation: science must give us an accurate trans-lation of the language of nature, because otherwise we could not inter-vene in the nonhuman world so successfully (134). Similarly, Hawkins cites with approval Richard Dawkins's simplistic argument that we should thank scientific realism whenever we fly an airplane and not a fly-ing carpet. If chaos theory works independent of culture, then so much for postmodernism, apparently (1995, 168). And Thomas Rice asserts that by revealing hidden order, chaos theory reverses postmodern skepti-cism and confirms the realist notion of independent reality (1997, 86).[23]

Yet Argyros himself says that "reality is not a fixed objective con-cept" and "the real is indeed a construction" (1991, 131, 134), and even that "there is no such thing as *the* world" (167). He goes on to state that chaos theory "appears to validate" Derrida's rejection of metaphysics' quest for eternal principles (256) and supports "a deconstructive descrip-tion of chaos in a sense" (257).[24] But he ultimately rejects deconstruction because he thinks we still need metaphysics to make technology work properly, and because he chooses to emphasize the creative, positive aspects of chaos rather than the destabilizing aspects. This seems to be a disagreement in emphasis, or a quarrel about what kind of metaphysics is needed to motivate scientific research.

Argyros does not rely only on chaos theory for his argument about realism, stating that "the evolutionary game presupposes a basically realist view of the physical, chemical, and biological worlds," which makes it, again, incompatible with deconstruction (105). Of course the practice of evolutionary biology or chaos theory (or any other science, for that matter) presupposes that researchers are studying the actual world rather than making stuff up. But this is far different from pre-supposing a metaphysical thesis about our access to mind-independent Reality—plenty of practicing scientists reject that metaphysical view. To the extent that deconstruction always seeks to show how systems undo themselves, and seeks to complicate our ability to refer to the world, this approach is doing something that most practicing scientists have no time for. In that sense, deconstruction is "incompatible" with any science, because it is awfully hard to do both deconstruction and chemistry at the same time. But then there is no need to cite the details of any partic-ular science to "disprove" deconstruction in this sloppy sense. Once the approach has been reduced to simplistic relativistic nihilism, even tying one's shoes or tripping on a rock can be put forward as "disproofs."

A number of other opponents of postmodernism have attempted to make similar use of chaos theory to defeat the supposed antirealism of deconstruction and other postmodern views. Hawkins identifies Derrida's slogan "There is nothing outside the text" with an antirealist view of science and states that "in arguments as it were diametrically opposed to those of some of the most influential modern literary theorists, the most influential chaos theorists everywhere acknowledge and indeed insist that there is 'something' of far more importance than their texts, outside their texts" (1995, 11). As if the straightforward realist attitude of everyday scientists could somehow provide a conclusive refutation of a metaphysical position about the relationship between language and reality. According to Hawkins, scientists "insist that their discoveries may represent increasingly accurate approximations of external reality in so far as they are tested, and may be overthrown, by negative evidence from nature" (72). And indeed, many do insist just this. But that insistence does not itself constitute a good argument for metaphysical realism, nor does it in any way rely on chaos theory.

Indeed, the fact that any other science could stand in for chaos theory in these arguments about realism raises the issue of distinctiveness: to what extent does Argyros make use of the special features of chaos theory in his borrowing? After all, he introduces the section of his book on chaos by stating that it "augments" the philosophical framework he has already presented (1991, 225), admitting that the main solutions to his central problem "have already been discussed at length" (228). Later he states that the concepts of chaos should be useful for "fleshing out" the framework already outlined (259) and that he seeks to translate his earlier conclusions "into the language of chaos" (290). In the case of the arguments against deconstruction, I contend that this translation adds nothing except the persuasive power of a more up-to-date science.

Deconstruction as Relativist. A realist position may seem attractive because it wards off the threat of relativism: if our science (or our morality, or our art) can reveal the actual nature of a world independent of us, then we need no longer worry that whatever people say is true (or good or beautiful) must be so. Hawkins, for example, identifies antirealism with extreme cultural relativism (1995, 36, 72), and Mackey also links these issues when he claims that "chaos theory refutes Hayles' postmodern faith in what might be called, ironically, an absolute subjectivism and relativism" (1999, 21).[25] Sokal and Bricmont even identify one of the key characteristics of postmodernism as "a cognitive and cultural relativism that regards science as nothing more than a 'narration,' a 'myth' or a social construction among many others" (1998, 1). In line with these concerns, Argyros seeks to cast deconstruction as inescapably relativistic, and he proposes that chaos theory's discovery of universals make it incompatible with postmodern thought.

According to Argyros, deconstruction can almost be defined by its "metaphysical conviction that nothing is universal" (1991, 281). Although admitting that such a stance may occasionally be useful for interpreting texts, he holds that when it comes to ethical or political choices, deconstruction is "powerless to make the most elementary of distinctions" because it views everything as context dependent with no universal basis for grounding choices (5). All that is left is "whimsy or expediency," leaving Derrida portrayed as a prisoner of paradigm relativism, which makes truth relative to a framework (93, 106).[26] And it gets worse: Argyros states that because of its theoretical failings, deconstruction is "doomed to serve the interests of stasis and reaction," and a theoretical system that cannot struggle for global good "is in de facto collaboration with terror" (75, 236).

Yet Argyros admits that Derrida renounces relativism and holds only that meaning is determined by context that is never absolutely fixed (1991, 30). He even agrees with Derrida that "context slippage" is unavoidable in human conceptual representations (188). But again these admissions do not stop Argyros from appealing to chaos theory's findings of "universality" in order to show that deconstruction conflicts with recent science (281, 340). Unfortunately, he confuses universality in the mathematical sense of Feigenbaum and others with the universality of aesthetic or ethical concepts, just as we saw with Hawkins in chapter 8.[27] Here again, attempts to exploit chaos theory in order to score philosophical points off Derrida run aground on uncharitable readings of deconstruction. Just as metaphors can suffer from mistakes about the source field, they can also founder on errors or exaggerations about the target.

Methods and Pluralism. As we have seen, some of Argyros's objections to deconstruction rest on misapprehensions and misconstruals. One of the most basic problems is that Argyros presents deconstruction as a "philosophical system" with a "basic premise" of fundamental instability (9) and speaks of its "fundamental axioms" (1991, 104). But he also recognizes it is as "essentially a negative methodology," and if it we think of deconstruction as an approach rather than a system then it becomes hard to see how chaos theory can be incompatible with its "principles" or "hypotheses" (34, 89). After all, a broad cultural orientation or flexible set of strategies does not present a tidy target for refutation. Along these lines, Knoespel points out an interesting comparison between deconstruction and chaos theory that deals not with their assumptions but with their strategies. Both start with traditionally neglected phenomena and bring new modes of investigation to bear on what had previously been avoided. They rely on heuristically useful examples rather than axiomatic schemes, and deconstruction goes even further by actively resisting becoming any kind of system at all (Knoespel 1991, 115–17). Now of course an antisystematic interpretive strategy will be "incompatible" with systematic scientific investigation, but this incompatibility hardly counts as evidence against deconstruction.

Argyros states that deconstruction's "most serious shortcoming" is its inability to "propose a positive agenda for either critical, aesthetic, or political action" (1991, 11), yet he recognizes that Derrida clearly states that deconstruction is not a "movement" with a "goal" that it can successfully achieve (60). The simple fact is that deconstruction seeks to do something different than what natural sciences (or many other kinds of inquiry) do. Just because deconstruction cannot do everything that we might want to do does not constitute a refutation of it. Argyros takes the example of an evolutionary theory of language, which emphasizes our continuity with the natural world, and suggests that this research project is "diametrically opposed" to deconstruction, which focuses on the distinctive meaning-making practices of humans (104). When Argyros complains that "deconstruction cannot ask 'Why?' or 'How?' in an evolutionary sense" (105), he seems to fault Derrida for not providing a biological theory of the origin of language, but why should that be the only appropriate goal or criterion? If we read Argyros's attacks on deconstruction as an attempt to redirect literary methodology, then it makes sense that he would invoke the natural sciences as alternative models for what kinds of questions to ask and what kinds of disorder to praise. This interpretation would enable a far more charitable reading of his book than if we conceive that the invocations of chaos theory are genuinely meant to serve as arguments that Derrida is wrong. A pluralist attitude would hold that natural science and literary interpretation are two endeavors pursuing different goals by different means. These endeavors may sometimes interact fruitfully, and sometimes they should leave each other alone.

Postmodernism in Legal Theory, Especially Critical Legal Studies

I have focused on literature and literary theory in this chapter, but postmodernism represents a broad cultural trend. For instance, postmodernism has affected legal scholarship, primarily in work known as critical legal studies (CLS). This postmodernist legal movement has also been linked to chaos theory; for example, law professor Bruce Arrigo describes chaos theory as one of the "leading approaches" in critical work connecting law and psychology—along with CLS, feminism, and postmodernism (2002, 151). Attorney Jay Moran has railed against deconstruction, postmodernism, and CLS for their relativistic "pollution" and "contamination" of legal education and scholarship (1997, 157). Moran goes on to state that promoters of CLS "have come to rely on chaos theory as an integral part of their postmodernist approach to law" (169). In this section I examine Moran's argument as well as the views of those who have found postmodernism (and its connection with chaos theory) less pernicious.

Moran's claim about the role of chaos theory in legal studies rests on thin evidence. Although he asserts that "postmodern intellectuals have often pointed to science as a key starting point in the development of their epistemes," he offers little more than a few general examples of this borrowing (1997, 172ff.). He mentions Lyotard's speculation about fractals, a citation that describes Derrida in terms of recent science, and a vague assertion that Foucault's reading of social science is equivalent to the revaluing of disorder in the physical sciences. None of this establishes a link with postmodern legal theory, although Moran does associate Prigogine's work with radical leftist critique because he views both as validating disorder.

Moran finds one concrete connection with the science of chaos in the discussions of Thomas Pynchon by practitioners of the "law and literature" approach. Now, we have already seen that Moran's treatment of Pynchon rests on a misapprehension of chaotic behavior as a merely senseless randomness. Moran's criticism of postmodern legal theory follows the same pattern as that of Argyros and other opponents of postmodernism: he claims that the practice of science requires some version of scientific realism, and then he excoriates his opponents for their antirealist position. Thus, Moran states that although contemporary intellectuals may be smitten with postmodernism, "the integrity of their kinship to science is less clear. If the breakthroughs in science are truly supportive of a postmodern era, one would expect these intellectuals to view modern science in much the same way that scientists do" (1997, 173). But one can borrow ideas from another field without importing the metaphysical baggage that supposedly accompanies them. Moran may be correct that scientists think chaos theory is just more business as usual (177), but even scientists may be wrong about the implications of their own work. Many scientists have described chaos theory as momentous and revolutionary, or even postmodern, whereas others reject all grand philosophical positions attached to their work, realist and antirealist alike. Yet Moran seems to think that we can summarily reject assertions of convergence between chaos theory and postmodern views just because some scientists who work on the former have expressed impatience or exasperation with the latter.

Moran makes a valuable observation that interpretive openness in law does not mean lack of pattern or structure, but he fails to forge a connection to the way chaos theory overcomes this dichotomy (1997, 195). Several other legal scholars have noticed this connection, however, which is usually expressed in terms of legal "indeterminacy." In fact, metaphorical restructuring has made a genuine contribution to jurisprudential discourse by suggesting a way beyond the fruitless conflict between conservative legal scholars and CLS theorists, which has often centered on issues of the "indeterminacy" of the law. On one side, partisans of the law and economics movement insist on the importance of predictable judicial outcomes, while those of the CLS movement typically hold that legal principles do not and cannot yield strictly predictable results. From the CLS perspective, legal decision-making is so fraught with biases in the service of existing power structures that it eventuates in random and indeterministic results whose only pattern is that the rich get richer (see, e.g., Bauman 1996, 1–4; Kelman 1987, 245–56). According to Denis Brion, legal indeterminacy can be defined as the situation in which "one cannot predict with a high degree of accuracy how the legal process will resolve particular disputes." CLS takes this indeterminacy to imply "that the law does not function autonomously of social, economic, and political power" (Brion 1995, 180).

Joan Williams characterizes some critical legal theorists as "irrationalists" because of their emphasis on indeterminacy. According to Williams, these scholars want to show that legal doctrine is so manipulable that "any side can use any doctrine for its own purpose. They argue that because law is not neutral, it is wholly indeterminate" (1987, 489). And from its indeterminacy they conclude that it is illegitimate and a disguise for domination. Williams draws an explicit connection to deconstruction when she observes that "the irrationalists, girded with Derridean learning, jump from the long-established tenet that law does not function by internal logic to the conclusion that law is therefore radically indeterminate, and that *any* legal argument can be argued for *any* given position" (491). Note that a metaphorical connection with chaos theory will suffer from lack of fit here because chaotic systems certainly do not lack an internal logic. But by now the reader will clearly spot a dichotomy ripe for being restructured by metaphorical borrowing, and I examine a number of attempts to do just that.

Andrew Hayes defends the integrity of the legal system against the all-out assault of those who insist on legal indeterminacy, but he rests his defense of legal principles on the image of the fractal, which he sees as a genuine boundary that is nonetheless not a straight line. He uses this image to challenge the view of legal indeterminacy, which says that legal rules are just a facade because any decision can go in any direction due to indeterminacy or the availability of counterrules. That critique assumes that "principles are straight lines with clear boundaries, and the only alternative is to confess that law is just a mask for free-form political struggle" (Hayes 1992, 766). Somehow, fractals serve as the dichotomy busters in this argument: "chaos theory shows that much, if not all, of the critical attack on jurisprudence rests on a false dichotomy; just as fractal shapes have complex borders rather than smooth edges, legal rules and doctrines also require careful examination to determine on which side of the jagged line a specific case lies" (766). Perhaps the fuzziness of fractal boundaries is here meant to call forth the exercise of principled yet undetermined judgment.

Roberto Corrada argues against postmodernism and CLS in favor of a unified theory of what is unconstitutional, based on what he sees as the benefits of unification found in science. He recognizes that postmodernist opponents of unified accounts may take comfort from chaos theory because "the seeming impossibility of predicting outcomes from chaotic systems may resemble the impossibility of deriving an explanatory principle that ties together the doctrine of unconstitutional conditions" (Corrada 1995, 1020). But according to Corrada, chaos theory actually gives us hope for unifying theories because of the universality of its results. Recurring patterns can be found in unpredictable and complex behavior, so "complexity should not deter legal scholars from seeking patterns in judicial outcomes despite the irregularity and unpredictability of the justifications used by judges in reasoning toward a particular end result" (1021). Even critics of the search for a unifying theory of constitutionality find evidence of recurring patterns in judicial decision-making, and "chaos theory suggests that these uniformities may be evidence of deeper patterns that exist beneath the thick layer of uncertainty that covers the surface. Complexity can ultimately be explained in a simple manner" (1027). And certainly it can be. But as we have seen, the "universality" found in chaos theory does nothing to guarantee that universally valid theories can be formulated for all messy phenomena. We still lack, for instance, any general theory of nonlinear behavior that would allow us to predict which systems will manifest chaos at specific parameter settings. As mentioned in chapter 1, chaos theory hardly counts as much of a theory in the traditional sense of a systematic account of its field of interest. So Corrada's attempt to harness the prestige of chaos theory to motivate a unificationist methodology works mainly by depriving his opponents of the same disciplinary resource.

Denis Brion also responds to the challenge posed by CLS, and he begins by examining a number of cases whose slight differences make different legal "themes" relevant for each one. Because there is no general agreement as to a hierarchical ordering of the values underlying these different themes, no general rule applies and case-by-case decisions are required. Although Brion characterizes this situation as "disorder" because simple appeal to legal doctrine yields indeterminacy, he insists that "we do not necessarily have caprice" (1995, 192). There is an underlying quasi-order at the level of theme, rather than doctrine, and "chaos provides a highly useful model for providing the descriptive terms of this quasi-order" (193). He spells out this suggestion as follows:

> The moments of judicial decision amount to the bifurcation points of chaos theory—the choice of outcome is not predetermined by an antecedent structure of legal rules. Corresponding to the strange attractors of chaos theory are the multiple, but nevertheless limited, themes of the law of tort—the paths from the bifurcation points lead to one or another of these themes. And ... a seemingly small difference in the facts acted as a Butterfly Effect, reversing the resolution of the case. (193)

Brion makes use of this metaphor (which more accurately describes an indeterministic system with multiple basins of attraction) to restructure a field organized by a dichotomy: "by analogizing to chaos theory as a means of description, we can see that structure is not the necessary condition of integrity; a disordered surface of legal doctrine does not thereby entail a law incapable of constraining what people do in seeking to achieve justice" (1995, 199). His argument suffers from a confusion of the chaotic with the merely disordered, but it nonetheless reaches a valuable conclusion: "surely the law in fact is indeterminate in part because some judges do impose their own ideology or ratify existing unjust distributions of power. This indeterminacy, however, is the result of the imperfection of human action. It is not a symptom of the impossibility of autonomy in the law" (199). Brion uses the order found in strange attractors as metaphor for intelligibility in indeterminate situations, and indeed good judgment cannot be reduced to a perfectly determinate algorithm. But concern for metaphorical distinctiveness leads me to ask: Do we really need chaos theory for this insight? And what does it mean if we require the disciplinary prestige of the sciences to help recuperate something that Aristotle said thousands of years ago? I return to these questions in the conclusion.

Conclusion: Pluralism and the Meaning of Science

My analysis of various arguments about postmodernism and chaos theory serves to emphasize the point made in chapters 5 and 6 about the importance of metaphorical fit in interdisciplinary borrowing. As discussed there, it is indeed important to "get the science right," since a mistaken use of a scientific theory or term may disrupt the effectiveness of a metaphorical borrowing. But at the same time, the mathematical structure of a theory does not fix its broader cultural meaning; a narrow focus on scientific correctness, such as we find in Gross and Levitt's *Higher Superstition*, runs the risk of missing out on the very real meanings attached to scientific results. Hayles (1990, 206) takes Michel Serres to task for getting his science wrong but finds herself excoriated for factual inaccuracies in her work (Gross and Levitt 1994, 101). It may seem that my criticisms of Hayles have much in common with the complaint of Gross and Levitt that Hayles's arguments are based on "subjective and shoddy analogies" and "leaky metaphors" (1994, 99). But they offer no criticism of her argument about chaos theory and deconstruction. Instead, they resort to abuse, and in doing so they seem to assume that a metaphorical argument requires no detailed refutation.

It is instructive to compare Gross and Levitt's treatment of Argyros with their treatment of Hayles. Argyros criticizes Derrida and other theorists for getting their science wrong (1991, 95, 227), but he himself makes a number of basic errors in his description of chaos theory.[28] These errors do not disqualify him from discussing the cultural implications of chaos theory, but they raise the question of how discussions of science are policed for accuracy, especially because Argyros himself engages in this police work when criticizing Derrida. Gross and Levitt argue that postmodern philosopher Steven Best is "ill equipped to draw inferences of any kind from his contemplation of chaos theory . . . simply because his grasp of these matters is so rudimentary and so tied to secondhand paraphrase." And after attacking Katherine Hayles for inaccuracies, they remark about Argyros that "his mathematical exposition is also far more systematic and coherent, although it is far from flawless" (Gross and Levitt 1994, 98, 270). In fact, the exposition in *A Blessed Rage for Order* makes a number of errors every bit as egregious as those cited in Hayles's work, and relies heavily on popularizations. More important, Argyros's comparisons between chaos theory and literary philosophy cannot possibly be "more cautious and tentative than those of Hayles," as Gross and Levitt state (1994, 270). After all, Hayles claimed only that a rough similarity between chaos theory and deconstruction shows that they both developed in response to some common cultural constraints. Argyros, on the other hand, claims that chaos theory shows Derrida to be wrong, and that nonlinear dynamics provides an objective ground for Enlightenment values, traditional narrative, and liberal democracy. These assertions by Argyros may accord with the hostility Gross and Levitt feel toward postmodernism, but they are not cautious claims.

One of the main lessons of this book is that metaphorical borrowings from the natural sciences should not merely be dismissed without argument. Any attempt to portray such cross-disciplinary transfers as simply misunderstandings of the science rules out the possibility of exploring the impact that science can have on the practices of other disciplines. Yet such cross-disciplinary forays are not meaningless—they can be undertaken and criticized and defended. Indeed, as mentioned in chapter 5, to condemn anything but literal use as unacceptable "misuse" would not clarify our language and eliminate muddle-headed mistakes. It would instead render communication nearly impossible.

When Gross and Levitt insist that "postmodern cultural transformation" is not "inscribed in the mathematical peculiarities of nonlinear dynamical systems" they are quite correct—there is no way straightforwardly to extract the cultural meaning of scientific research from the equations in the textbooks (1994, 105). Hayles herself criticizes Lyotard for his breezy recruitment of chaos theory as validation of postmodern values because his claim "is akin to social Darwinism, in that it confuses scientific theories with social programs" (1990, 215). As we saw in chapter 7, evaluative questions cannot be collapsed into scientific ones, nor replaced by them, nor settled by a breezy leap from facts to values. Yet this does not imply that cultural meanings drawn from science are merely subjective or idiosyncratic. Such a view belies an inability or unwillingness to see the very real cultural meanings of science.

One of the best examples of this issue is the early twentieth-century response to the Special Theory of Relativity. It was popular to draw an easy conclusion from Einstein's work: physics has proven that there are no absolutes and "everything is relative." This glib connection between Relativity and relativism came under fire from scientists who insisted that this popular interpretation of Einstein's theory ignored the details of the physics. The Special Theory did not make everything "relative"; instead, it proposed a new geometrical structure for mechanics. This four-dimensional structure had no shortage of objectively correct results and absolutely precluded possibilities. In fact, Einstein thought about calling his work the Theory of Absoluta or the Theory of Invariants, because it provided a four-dimensional vectorial measure that was precisely *not* relative to an observer's inertial frame (Hentschel 1990, 178). But "relativity" won out.

And perhaps this is as it should be. Scientists who protest loudly of popular misunderstandings of relativity may be so narrowly focused on the mathematical structure of the theory that they ignore the fact that a theory is more than just that structure. A theory carries with it a way of doing science, a way of looking at the world, one might even say a "culture." Newtonian science with its absolute space and time had profound cultural resonances, so it is only right that a theory that did away with these absolutes should also have profound cultural effects. Whether or not we can mathematically transform Newton into a "relativistic" framework or reconceive Einstein as providing universal "invariants" does not eliminate the broader cultural aspects of scientific results. The simple fact is that people felt Einstein pulling the rug out from under them, and they were not wrong to feel this way. Relativity does indeed have *something* to do with relativism.

The situation of the Heisenberg uncertainty principle parallels that of relativity. It has become commonplace to invoke Heisenberg in discussions of anthropological observation or journalistic objectivity: the physics of subatomic particles supposedly proves that any attempt to observe an event changes that event. This metaphorical extension from quantum mechanics to social situations faces a fundamental challenge: the mathematical structure that underlies the uncertainty relations simply makes no reference to observers, or consciousness, or subjectivity. Yet the persistence of philosophical quandaries about measurement in quantum systems makes it clear that the popular understanding of Heisenberg is not entirely outlandish. Physicists had been one source of a view of knowledge as ideally removing the observer from all responsibility for the situation in which knowledge is created. In quantum mechanics, physicists encountered a situation where such a separation of the knower from the known was highly problematic. The assimilation

of this difficulty by the broader culture, with all its attendant simplifications and misunderstandings, nonetheless contains an appropriate
response to this news about scientific knowledge.

Similarly, although it is tempting to chastise writers who draw momentous conclusions from chaos theory, it is not enough simply to point
out that unpredictable plot twists are not the same as sensitive dependence, or that we have no conclusive evidence of low-dimensional chaos
in complex social systems. As Hayles explains, the long history of cultural traditions encoded into a word like "chaos" mean that "chaos
theory can scarcely avoid having implications for the culture beyond its
technical achievements" (1991, 3). Gross and Levitt insist that the mathematical concept of linearity has nothing whatsoever to do with cultural
questions about instrumentality or pluralism, but they are incorrect. To
the extent that our widely accepted frameworks—even frameworks for
evaluative inquiry about human society or art or religion—have been influenced by the physics of regular linear behavior, those frameworks can
indeed be challenged by methodological reorientations and metaphorical antidotes drawn from chaos theory. Protesting that political meanings of a science are not visible to the practitioners themselves in no way
removes the potency of those meanings (contra Porter 1991, 1582).
Human actions and utterances have unintended results, but this fact
does not render the unintended meanings and effects merely subjective.
And the meaning of the knowledge produced by a discipline cannot be
entirely fixed within its own bounds (see Moran 2002, 163).

Although I have examined borrowings *from* the natural sciences, influence runs not only in one direction. Paul Forman (1971) has argued
that the development of quantum mechanics must be understood in the
cultural context of Weimar Germany, for instance. Although nonscientists appropriate scientific ideas and terminology, scientists also use
conceptual resources from the nonscientific cultures of politics, art, and
philosophy.[29] Because science is part of this broader cultural context,
no one group of people has a preemptive right to decide what chaos theory does or does not mean. The discussions of the relationship between
chaos and economics, law, or literature constitute part of the process by
which such meanings are created. Of course, one might argue that our
frameworks for thinking about human society or art should never have
been so influenced by the results of physics. That is, if we hadn't been
so zealous in our pursuit of knowledge modeled on linear physics, then
new results from physics would never have seemed so momentous.

: : :

Chaos theory has proven a fruitful arena for investigating the interaction between disciplines. Sokal and Bricmont display an admirable attitude of disciplinary pluralism when they say that "the type of approach in each domain of research should depend upon the specific phenomena under investigation" (1998, 188). So, for example, psychologists should not need to justify the effect of observation by appeal to Heisenberg. But they then proceed to say that because there is a lot the natural sciences still do not understand, "it is perfectly legitimate to turn to intuition or literature in order to obtain some kind of nonscientific understanding of those aspects of human experience that cannot, at least at present, be tackled more rigorously" (188). Note the condescension in treating literature as a second-best or stopgap measure, useful until we can get around to serious inquiry. The interaction between disciplines requires a willingness to investigate the broader implications of scientific research as well as mutual respect.

This issue of respect highlights the fact that the investigation of borrowed knowledge raises questions not only about who decides the meaning of the latest research in the natural sciences but also about who decides other things as well. Scientific knowledge cannot immediately decide pressing evaluative questions about art or society, but neither is it irrelevant to aesthetic and ethical inquiry. Trained scientists are not the only ones qualified to discuss the meaning of science, and English professors are not the only ones who should be allowed to discuss the meaning of novels. Gross and Levitt state that anyone who attempts a "serious investigation of the interplay of cultural and social factors with the workings of scientific research in a given field . . . must be, *inter alia*, a scientist of professional competence, or nearly so" (1994, 235), but this assertion considerably restricts the opportunities for critical discussions of cross-disciplinary borrowing. By contrast, Sokal and Bricmont promote the pluralist value of open participation when they aver, "we have no desire to prevent anyone from speaking about anything. . . .The intellectual value of an intervention is determined by its content, not by the identity of the speaker, much less by his or her diplomas" (1998, 12). The pluralist approach has important implications not only for the relationship between disciplines, but for broader questions about participation in knowledge making and decision making. The examination of borrowed knowledge calls for the virtues of genuinely open-minded critical conversation that pluralism commends.

Although chaos theory may have faded from scientific fashion, the phenomenon of borrowing will become increasingly important as new discoveries in the natural sciences continue to excite broad scholarly and popular interest. The tremendous resources devoted to medical science will influence our concepts of health and the human. Research into the function of the brain will call forth broader applications from everyone who has something to say about thinking. And soon we will see the rise of the Next Big Thing in science—perhaps a breakthrough in proteomics, or quantum computing, or the discovery of the provocatively named "God particle." I have no predictions to offer about what it will be, but I am sure that researchers in the social sciences and the humanities will seek to borrow it for their own endeavors. When that time comes, the tools developed in this book will be available to help sort through it all. And this investigation into borrowed knowledge has also revealed some important lessons for those who would "lend" their knowledge across disciplinary lines. Such lending should not be rejected out of hand as unhelpful, for although some offers of assistance disguise imperialistic motives, not all of them do. The message of pluralism here is that disciplines can sometimes interact fruitfully. In some cases, a field will need to fight off unwanted and unhelpful attempts to lend concepts and methods, while in other circumstances a discipline may even turn out to be a good candidate for subsumption and absorption. Pluralism counsels an open-minded, empirical, case-by-case approach to lending as well as borrowing.

The case of chaos theory as source for borrowed knowledge yields lessons not only about the implications of science but also about our knowledge-making practices more generally. If we accept that each of the different disciplines offers valuable approaches that can sometimes inform and influence others, then we can learn these broader lessons of cross-disciplinary borrowing. If we conceive of knowledge production as a social process, we may be able to stop separating off "persuasion" and "rhetoric" as threatening contaminants and instead investigate the risks and advantages of specific persuasive techniques. If we recognize that metaphors can serve a valuable conceptual function, we can develop and refine criteria for critically examining them. If we move beyond the sterile opposition between supposedly "value-free" facts and supposedly subjective, irrational values, we can examine argumentative bridges between them for faults and benefits. Our practices of inquiry may then come to exemplify the pluralist virtues of open-mindedness, humility, responsibility, and respect.

Notes

CHAPTER I

1. See, for example, Psalms 111 and 118, and the Recitation of Sins from the Yom Kippur liturgy; more recently, I take my inspiration from Ian Hacking (1999), who compiled an acrostic list of things that have been claimed to be "socially constructed."

2. My colleague David Davies informs me that these dynamics are also illustrated by the stretching and folding involved in the making of traditional *la mian* noodles in China.

3. Nonlinear equations include formulas with cross-terms as well as those with variables raised to powers other than one. Carl Chiarella and Peter Flaschel (2000, 141) find complex dynamical behavior, including chaos, in a standard Keynesian model of monetary growth without requiring nonlinear economic behavioral relationships such as nonlinear functions for investments or demand. Instead, their model simply has some of the economic relationships linked in such a way that one variable is the product of two others.

4. A dynamical system is a well-defined mathematical entity. Just because something changes in time does not automatically qualify it as a dynamical system in the strict sense (contra Argyros 1991, 114).

5. Describing what he views as the inexorable march of probabilistic concepts through physics, Philip Mirowski declares that "chaos theory intoned the funeral rites by summoning the 'fractal geometry of nature' to explain the futility of Hamiltonian dynamics" (1989, 74). But chaos theory explains no such thing. Hamiltonian dynamics—that is, the classical

nineteenth-century physics of systems that obey conservation of energy—is still fruitful and useful for many purposes. Indeed, much research in the chaotic behavior of systems without friction (such as the solar system or subatomic particles in an accelerator) relies crucially on Hamiltonian dynamics.

6. N. Katherine Hayles seems to display some frustration when she intones that "fractal geometry is not the same as nonlinear dynamics; nonlinear dynamics is not the same as information theory; information theory is not the same as thermodynamics" (1990, 292). It is hard not to sympathize.

7. The description of attractors as exerting a "magnetic" appeal for the system may result from the popular work of Briggs and Peat (1989, 36). I can assure the reader that it is not the result of the source misquoted by Bruce Arrigo, who describes attractors as encouraging disorder by saying, "behavior governed by this magnetic pull is called strange or 'butterfly' attraction (Kellert, 1994)" (2002, 162).

8. For example, a symposium invited eighteen constitutional scholars each to trace the consequences of one historical change in law, and the proceedings bore the title "The Sound of Legal Thunder: The Chaotic Consequences of Crushing Constitutional Butterflies" in homage both to chaos theory and the famous short story by Ray Bradbury (Chen 1999).

9. I will generally be examining books that were published by major university presses.

10. The list includes Hayles (1991, 22), Hayes (1992, 762), Conte (2002, 18), and McCarthy (2006, 42), as well as the references cited in Kellert (1992, 39).

11. For a discussion of the difficulties in applying these techniques, see Kellert (1995). Consider also the advice of Sokal and Bricmont: to apply chaos theory "in a sensible way, one must first have some idea of the relevant variables and of the type of evolution they obey. Unfortunately it is often difficult to find a mathematical model that is sufficiently simple to be analyzable and yet adequately describes the objects being considered" (1998, 145). See also van Peer (1997, 46) and Pigliucci (2000, 64). In the humanities, attempts to apply the quantitative techniques of chaos theory are severely problematic. For example, Kundert-Gibbs (1999, 154) attempts to produce a phase-space diagram of Samuel Beckett's play *Footfalls*. See also Greeley (1995).

12. But see Cardim de Carvalho's article on economic themes in Shakespeare (2003), which, by the way, mentions the nonlinear nature of history and the way small events (such as Antony's eulogy in *Julius Caesar*) can have momentous consequences (218).

13. There have been some quantitative studies of the law using techniques from chaos theory. Post and Eisen used a database of cases decided by New York Court of Appeals and the Seventh Circuit Court, and they found some support for the notion that common law decision-making exhibits fractal structure: "The fractal structure hypothesis accounts for a limited (though statistically significant) fraction of the variation in citation practices" (2000, 577). See also DiLorenzo (1994, 435).

14. Clark's talk of "lawlike generalizations," together with his statement that the goal of science is "reasonably complete, covering-law explanations,"

will strike many as hopelessly old-fashioned. But Clark dismissed criticisms of these pre-Kuhnian notions as merely "intramural disputes" among philosophers of science (1981, 1238).

15. In our critical examinations of the arguments found in published articles of legal research, we should also bear in mind that academic journals differ, with respect to their reviewing processes and standards, from field to field and from journal to journal. Because of this great variety, one may occasionally find a published essay that declares, "this Article is not, however, yet another ponderous self-serious article on theoretical jurisprudence. It is about life, law, discovery, and play. Herein I speculate on the speculative and allow myself the recreation of dabbling, comparing, and contrasting many disciplines of study" (Geu 1994, 935). When an author presents material as playful speculation, critical analysis may be an inappropriate response.

16. One might also include "Strange Attractors," a short story by Rebecca Goldstein (1993).

17. Gordon Slethaug cites William Carlos Williams on the influence of Einstein on poetry and a number of other scholars on the influence of quantum theory and relativity on the form of the modern novel (2000, 3–6). Mackey speculates that Joyce may have read (or read about) Poincaré while in Paris (1999, 58–59).

18. With regard to the relationship of physics to economics, Mirowski sometimes says that changes in the two fields coincided not because of base-superstructure determination but because of "resonances" between the metaphors in different fields, much like Prigogine's talk of a "common intellectual climate" (Mirowski 1989, 119; Prigogine 1993, 5). At other points, Mirowski says conceptions borrowed from physical science "channeled the evolution of economic theory" (1989, 192) or that dominant theories of value have actually been "dictated by the evolution of physical theory" (396).

19. Weinberg (1996) makes exactly two exceptions to this claim: the cultural impact of the discovery of mathematical laws of nature and the impact on philosophy when it finds that some of its subject matter belongs instead to science. Tellingly, he leaves aside metaphorical uses almost as an afterthought.

20. The pluralism I advocate has connections with value pluralism, religious pluralism, and other varieties of philosophical pluralism. See, for example, the essays in Archard (1996). I regret that I am unable to explore these connections here.

CHAPTER 2

1. These levels should not be thought of as rigidly walled off from each other. After all, physicists and economists themselves often ask and answer level-2 questions. Indeed, sometimes their work at level 1 is precisely in service to a defense of a favored philosophical or methodological point. For example, economists' analysis of a particular commodity market (a level-1 enterprise) might be primarily motivated by a desire to demonstrate the superiority of their preferred data-analysis technique (a level-2 concern). Furthermore, our level-1 accounts of atoms and markets color what we will look for—and what we will

see—on level 0, just as our methodological commitments on level 2 will affect which theories we choose on level 1.

2. On this point, see also Nozick (1981, 644).

3. Thanks to Geoffrey Gorham for helping to raise this issue.

4. This danger is raised by Bechtel (1987, 298), Hansson (1999, 340), and Bauer (1990, 113).

5. Thanks to Sara Mack for suggesting this metaphor.

6. Apart from these three kinds of images, some writers characterize the relationship between disciplines by borrowing concepts from the sciences (Klein 1990, 80). For instance, we often find economic images of markets, with competition and entrepreneurs, importation and trade. Economic metaphors also appear in the work of historian of economics Philip Mirowski, but with considerably less savory connotations. He describes the relations between the founders of neoclassical economics and mid-nineteenth-century physics as "wholesale piracy," "brazen daylight robbery," and freeloading (Mirowski 1989, 5, 9, 109). Biological metaphors include Levins and Lewontin's notion of a science surrounded by a semipermeable membrane that can admit or stop traffic in or out as well as conceptions of disciplines as organisms with hybrid vigor, cross-fertilization, and symbiosis (Klein 1990, 80). Novelist John Barth also favors biological metaphors when he describes how chaos theory is taken up by various sciences and "comes to infect literary theory as well" because it is "an idea too rich, a metaphor too powerful, not to spread 'rhizomatically' out of its original bounds into other fields, like crabgrass on a suburban American lawn" (1995, 284). Literary scholar Thomas Jackson Rice favors images from physics, such as the conception of disciplines as parts of a vector field in which each one affects and is affected by every other, or the notion of a "red shift" time lag between the ascendance of a scientific theory and its assimilation by literary scholars (Rice 1997, 5, citing Hayles 1984, 109). We can also find psychological images such as rivalries between siblings and cybernetic images of systems with feedback.

7. The analogue to interfield theories, that is, the creation of a new nation between two existing ones without changing the original two, is a mismatch for this image.

8. Perhaps this situation is also the case with genetic and environmental causes of a particular disease. See Longino (2006).

9. The Monte Carlo procedure is a technique for estimating the solution to an intractable problem by simulating a number of trials with random numbers.

10. This discussion relies on Nickles (1980) and Hoyningen-Huene (1987). I see now that several of my points connecting the rhetoric of inquiry and the context of discovery were anticipated in Nickles (1998).

11. These three related dimensions of the two-context distinction do not line up neatly. The temporal dimension is clearly different from the others, and the empirical/logical and descriptive/normative dimensions are closely related. See Nickles (1980, 8) and Hoyningen-Huene (1987, 504–6). A descriptive account of an activity may very well include a description of the norms that govern that activity, but this fact will not make it normative in the sense used here.

12. This conception of the a priori provides an additional point of contact between the logical and the empirical, further undercutting that aspect of the two-contexts distinction.

13. For a spirited objection to the naturalistic treatment of the a priori, see Friedman (1997).

CHAPTER 3

1. Notice that here, as elsewhere, I rely on the rhetorical strategy of characterizing my own position as the moderate one, intermediate between extremes.

2. See Longino (2002) on the mistaken opposition between the social and the epistemic.

3. Thanks to Alan Gross for reminding me of this point.

4. This division draws upon some recent work on relationships between economics and the natural sciences, which offers even more detailed taxonomies of these purposes. See I. B. Cohen (1993, 13) and Niman (1994, 361–62).

5. For fractals, see Livingston (1997, 20, 34, 56, 118, 147, 166). For strange attractors, see Livingston (1997, 29, 60, 130). And lest it be thought that only English professors flavor their prose with chaos theory, consider Paul Carrington's address, published in the *Duke Law Journal*, which invokes the butterfly effect to illustrate a point about the modest power of legal education. His discussion of sensitive dependence serves as little more than a colorful opening anecdote (1992, 742–43).

6. Logic texts call this the fallacy of denying the antecedent.

7. For claims of anticipation in romanticism, see Barth (1995, 328); for Nietzsche, see McCarthy (2006, 134); for culture in general, see Demastes (1998, 11); for a claim that several postmodern novels "anticipate developments in the science of orderly disorder," see Conte (2002, ix). Later, however, Conte retreats from this claim: "it would be inappropriate to say that these novels 'anticipate' discoveries in the fields of fractal geometry or nonequilibrium thermodynamics" (2).

8. Hayles offers a different interpretation of the proposed convergence of chaos theory with preexisting wisdom, acknowledging that "it may appear that the new science is simply discovering what everyone has known all along" (1990, 144). Yet she argues that chaos theory gives classic, romantic, and modernist conceptions of the relationship between order and disorder a subtle new twist. A change in focus and emphasis, she points out, is a real change in knowledge.

9. I use the term "intellectual authority," as opposed to cognitive or epistemic authority, to signal that what is meant is the perceived right to participate in intellectual debates. Scientists enjoy the presumption that their contributions are relevant and worthy of consideration, whereas nonacademics and practitioners of other disciplines may have to fight to be heard in the same forum. Helen Longino makes the useful distinction between cognitive authority, which is gained by training and ought to be distributed based on actual expertise, and intellectual authority, which ought to be distributed equally (2002, 133).

10. Gross and Levitt seem to approve of this kind of foray into the humanities (1994, 269). I do not disagree.

11. My own rhetorical strategy, which involves large numbers of quotations, may seem to rely on both the recruitment of allies and the "more is better" approach that rhetoric scholars call the topic of quality.

12. Klein, following Pomerance, notes that among the common problems with borrowing are the "'illusions of certainty' about phenomena treated with caution or skepticism in their original disciplines" (1990, 88). The prestige of the natural sciences can allow one to seem authoritative precisely because one has switched to a new context in which the hedges and qualifications of the original discipline may drop away. With nonlinear dynamics, it has often been the most vocal boosters of chaos theory who get cited by those doing the borrowing, rather than the doubters. The difference is even more salient in the case of so-called complexity theory.

13. Given some historical chasms between the sciences and literary study, the invocation of the natural sciences might not carry with it any prestige in certain contexts. Some old-fashioned humanists committed to disciplinary isolation may well scoff at any attempt to enroll scientists as allies.

14. In some of the fields under consideration, virtually anything is a fair candidate for pursuit. When one doesn't need expensive equipment and the empirical constraints on speculation are distant at best, evaluating hypotheses for pursuit may sometimes simply collapse into the question of audibility: "What can get me published?"

CHAPTER 4

1. We can see an early example of this methodological appeal in Immanuel Kant's suggestion to reform metaphysics on the model of the revolutionary triumphs in mathematics and natural science. The great success of these latter disciplines "should incline us, at least by way of experiment, to imitate their procedure, so far as the analogy which, as species of rational knowledge, they bear to metaphysics may permit" (Kant [1787] 1929, 22 Bxvi).

2. Samuelson invokes a contrast between regularity and randomness in his discussion of business cycles: "No exact formula, such as might apply to the motions of the moon or of a simple pendulum, can be used to predict the timing of future (or past) business cycles. Rather, in their rough appearance and irregularities, business cycles more closely resemble the fluctuations of disease epidemics, the weather, or a sick child's temperature" (Samuelson 1980, 237). Oddly enough, early work on chaos theory studied both epidemics and the weather. It is interesting to speculate on why Samuelson was not among the first to examine chaotic models of the economy, given that he was trained in the dynamical tradition of two of the founders of nonlinear dynamics, Poincaré and Birkhoff (Weintraub 1991).

3. A standard source for this approach is Lucas (1975).

4. This model shows that complicated and unpredictable behavior can occur as a result of the rational pursuit of cost-effective predictions. A similarly ironic result appears in a model that shows how the inability of economic agents to

distinguish chaos from randomness can itself cause chaotic behavior in the model (Sorger 1998, 380). Another interesting model demonstrating chaotic behavior resulting from the diversity of economic agents can be found in the recent work of Onozaki, Sieg, and Yokoo (2003).

5. As Brock and Malliaris explain it, nonlinearity can show up as temporal dependence in the residuals of the best linear fit, after deseasonalizing and detrending (1989, 301). They believe it may be plausible to think of long-term nonlinear structure in stock returns as linked with chaos in the business cycle (334).

6. Many of the searches for chaotic behavior use the BDS test, which examines evidence for low-dimensional attractors by measuring the correlation dimension of a data set. The original unpublished version is by Brock, Dechert, and Scheinkman (1986); the published version is by Brock et al. (1996).

7. Randall Bausor makes some interesting observations about much of the early work on chaos in economics. He suggests that most of the investigation has sought to establish the presence of chaos, rather than looking at the nature of phase transitions in these systems (1994, 119). With regard to empirical research on economic chaos, Bausor suggests that the techniques of nonlinear dynamics require analysis of the local behavior of the stable and unstable manifolds near critical points, "and this cannot be fulfilled without reference to the equations of motion" (121). But several economists have in fact looked at ways to reconstruct a system's dynamics without access to its underlying equations and have made use of bifurcation theory: Grandmont (1989, 46); Foley (1989, 93); and Barnett and Choi (1989, 143). Bausor is correct, however, that the techniques used to find the dimension or Lyapunov exponent of a strange attractor in price data can never find bifurcations, because they presume that they are looking at a structurally stable system (1994, 120). See also Barnett and Hinich (1993, 256).

8. Mirowski's argument is a bit technical, and I deal with it in greater detail in later chapters. I take his main points to include the following: that the neoclassicals chose to define market price as the price that clears the market in a single moment, thus enshrining the physics of equilibrium and making price dynamics exceedingly difficult to conceptualize (Mirowski 1989, 240). Furthermore, the founders of neoclassical economics drew on an analogy of value as energy, and constrained maximization of utility as equivalent to the variational principles of dynamics (Mirowski 1989, 219–21). But earlier economists such as Pareto did not understand the physics, so they did not see that if utility is an integrable, conservative field, then expenditure (kinetic energy) + utility (potential energy) is a conserved quantity (Mirowski 1989, 247).

9. Curiously enough, those economists who have embraced chaos theory are berated by Philip Mirowski for seeking to save a deterministic, mechanistic approach rather than wholeheartedly welcoming the stochastic nature of economic interactions. Indeed, chaos theory is strictly deterministic, and Mirowski may be correct that the incorporation of chaos theory does not represent a fundamental revolution in economic theory. As I have argued elsewhere, momentous *methodological* change does not always go hand in hand with challenges to fundamental *theories* (see Kellert 1992; chap. 1 of this vol.).

10. Neil Niman makes a similar point about motivating methodological change in economics when he contends that the real significance of borrowing models from evolutionary biology is that they can promote a shift from an approach that seeks prediction based on fictional constructs to one more solidly rooted in factual description (1994, 380).

11. I regret that my focus on economics, law, and literature does not permit a fuller discussion of implications for other fields. For example, Charles Dyke (1990) uses nonlinear dynamics to make a powerful methodological argument regarding history: momentous events do not require proportionately great causes. See also McCloskey (1991, 31–32) and Lindenfeld (1999, 287). For a fuller discussion of this and related points, see Kellert (1995, 39–42).

12. Perhaps the most poetic invitation to relinquish certainty in the study of law comes from Milner S. Ball (1985, 15), who before the advent of borrowings from chaos theory cited James Boyd White (1984, 278): "When we discover that we have in this world no earth or rock to stand or walk upon but only shifting sea and sky and wind, the mature response is not to lament the loss of fixity but to learn to sail."

13. We should be careful here not to conflate the strange-attractor "school" of chaos theory with the work of Ilya Prigogine on "dissipative structures." Recall the discussion in chapter 1. Strange attractors represent a challenge to economic equilibrium by expanding the notion beyond steady states. Prigogine's work attempts a much broader challenge to what he sees as the longstanding focus on equilibrium in all of science. The former studies new types of dynamical behavior that some economists see as legitimate instances of economic equilibrium. The latter looks at systems "far from equilibrium."

14. In contrast, sometimes knowledge from the natural sciences is harnessed for the purpose of undermining the notion of scientific objectivity. Quantum mechanics and relativity theory often serve as the source for such arguments. I owe this observation to Mike Reynolds, and I consider some arguments of this type in chapter 9 on postmodernism.

15. This is an example of the "antidote" function of metaphorical borrowing, which I discuss in the next chapter.

CHAPTER 5

1. Terminology varies. Instead of Lakoff and Johnson's "source" and "target" (1980), Perelman and Olbrechts-Tyteca speak of the phoros and the theme (1969), whereas others prefer to speak of a primary and secondary element.

2. As Ben Denkinger has pointed out to me, no field worthy of the name lacks all structure entirely. But some of the elements of a field may not yet stand in specified relationships.

3. The relationship between metaphors and analogies has been characterized in a number of different ways. Dedre Gentner and Michael Jeziorski consider metaphor to be the broader category and conceive of analogy as a subcategory concerned only with relational matching (1993, 449). Some have sought to draw an invidious distinction. For instance I. B. Cohen reinscribes the difference between the genuine and the "merely rhetorical" into a distinction between

analogies and metaphor, with the latter serving only as ornamentation that attracts attention or seems "scientific" (1993, 38). My discussion makes use of the interaction view to analyze some analogies in terms of metaphorical function, but I agree with Kittay that metaphorical meanings are more than just far-fetched analogical extensions. They retain both semantic fields active in interaction, whereas nonmetaphorical analogies do not bring over new content from the source field and remain symmetrical: the tornadoes of Kansas are like the hurricanes of Florida (Kittay 1987, 151–52).

4. Thanks to Jordan Macknick for raising this point.

5. Among those who talk of isomorphism, see Black (1993, 30).

6. Many of these uses of metaphor are mentioned in Klein (1990, 93). See also Hesse (1966) and Haack (1998, 83).

7. Unsurprisingly, Mirowski disagrees with this positive assessment.

8. Milner S. Ball touches on both the structuring and restructuring capacities of metaphor when he notes that conceptual metaphors for law can "circulate, diversify, increase, stimulate the creating of other metaphors, and challenge the hegemony of monolithic conceptual thinking" (1985, 17).

9. Arjo Klamer and Thomas Leonard suggest that the suspicion of metaphor among economists also rests on the supposition that it introduces unwanted ambiguities (1994, 20).

10. Robert Artigiani makes a similar claim about translating scientific paradigms to the study of history (1992, 132). Perhaps the ultimate example of this insistence on the literal comes from cognitive scientist Zenon Pylyshyn. He admits that the metaphor of brain as computer and mind as software is widespread. But he holds that computation, as the rule-governed manipulation of symbolic representations, "applies just as literally to mental activity as it does to the activity of digital computers" (Pylyshyn 1993, 557). We have encountered live metaphors (which still retain their novelty and productivity) and dead metaphors (which have passed into common use as new technical terms). Here, however, is a metaphor that has been killed and reanimated as one of the undead.

11. My point here is quite different from the suggestion of Weingart and Maasen that the metaphorical transfer of knowledge between disciplines is itself a deterministic nonlinear process (1997, 514–20). In effect, they propose the use of chaos theory itself to characterize the borrowing of chaos theory.

CHAPTER 6

1. Our willingness to take seriously the thesis that there are no unfunny jokes may explain why philosophers are not considered to be fun at parties.

2. In his essay "Do Metaphors Affect Economic Theory?" Maurice Lagueux suggests that although metaphors can be "evaluated on various grounds (aptness, relevance, clarity, insightfulness, etc.)," such evaluations "tend to be overshadowed by considerations related to rhetoric (are they persuasive?) and to literary quality (are they good stories?)" (1999, 3). I seek to avoid allowing this overshadowing to happen, and—as discussed in chapter 3—I do not consider rhetorical considerations to be entirely isolated from epistemic ones.

3. My examples of metaphorical borrowing come mostly from the fields of law and literature, because the use of chaos theory in these fields is more clearly metaphorical than in economics. Lagueux claims that metaphorical terms in economic theory are either dead metaphors simply used as convenient nomenclature or else "they are perfectly dispensable" tools for explication and persuasion (1999, 17). Yet his rejection of any substantive role for metaphor in economics seems to rest on his loyalty to a strict distinction between the context of discovery and the context of justification (16–17), for he admits that metaphorical borrowing can put a theory in a position to "raise many new questions" that would not have otherwise been asked (18).

4. One approach in the discipline of cognitive science is to measure the degree of fit in terms of whether the source and target occupy similar locations in their respective cognitive "spaces." Some have used this technique to measure fit as a way of determining the degree of comprehensibility (and even the degree of "aesthetic pleasingness") of metaphors. On this approach, "An ICBM is a haystack" fails in comprehensibility because of the placement of the two terms when their cognitive fields are superimposed: "their locations in these subspaces are quite discrepant" (Sternberg, Tourangeau, and Nigro 1993, 285). Such a measure may capture one aspect of fit, but it is limited to examining only locations within superimposed cognitive subspaces, without paying attention to the quality or nature of the relationships between terms in each field.

5. On Darwin, see Gordon (1989), as cited in Schabas (1994, 323). On Freud, see Grattan-Guinness (1994).

6. Mirowski seems to acknowledge this point to some extent when he states: "The indictment that some historical figure was 'mistaken' in appropriating a metaphor from another discipline is always a difficult case to prosecute, if only because a metaphor need not bear an identical resemblance to the initial object of comparison in each and every respect" (1989, 228). This highlights the trade-off between fit and utility that is discussed in the next section.

7. But note that Barondes himself makes mistakes in his presentation of the science of chaos, claiming that a chaotic system is one in which small changes in *parameters* have large effects (1995, 171).

8. Jeffrey Rudd (2005) has recently subjected Ruhl's borrowings from chaos and complexity theory to serious and detailed criticism with regard to fit. He finds a number of mismatches between sociolegal systems and the natural systems studied by evolutionary biology but concludes that many of Ruhl's proposals for regulatory reform have merits that do not depend on the analogies cited.

9. These examples of sensitive dependence might more strictly count as instances of instability, but they are relatively straightforward. I do not claim to be able to evaluate William Demastes's assertion that "the butterfly effect is in fact a scientific counterpart to Solness's trolls" in Ibsen's *The Master Builder* (1998, 84). And Slethaug certainly overstates the case when he claims that any appearance of an effect disproportionate to its cause is "a sure sign of the use of chaos theory in fiction" (2000, 18).

10. Reynolds unfortunately goes on to cite Leff: "If a state of affairs is the product of n variables, and you have knowledge of or control over less than n variables, if you think you know what's going to happen when you vary 'your'

variables, you're a booby" (Leff 1974, 476). Oddly enough, work in nonlinear dynamics shows this to be false: on at least some occasions, control over a small number of variables can in fact enable robust predictions. Reynolds has confused unpredictability resulting from sensitive dependence with unpredictability resulting from complexity.

11. Thom certainly has the right to protest the misuse of mathematical terminology, given some of the kooky extrapolations made from his original work on catastrophe theory. Note that not only mathematicians warn of this kind of misuse. Hayles mentions "the equivocations that result when a concept is imported from one discipline to another," especially in the work of Michel Serres (1990, 26, 198), and I. B. Cohen cites Freud on the danger with concepts when one undertakes "to tear them from the sphere in which they have originated and been evolved" (1993, 14).

12. Harriet Hawkins forthrightly admits that she uses the words "chaos" and "turbulence" interchangeably, even though they have different technical meanings (1995, x).

13. It is true that strange attractors occur only in dissipative systems, but chaotic behavior occurs in a broader set of systems than those that can have attractors. David Kelsey acknowledges the difficulty of applying chaotic models of dissipative systems in economics, where there is no real analogue of energy or friction, although some processes can reduce the dimension of the dynamics so that the analytical techniques of low-dimensional attractors might be applicable (1988, 25), and Mirowski cites Kelsey in observing that one cannot have strange attractors in Hamiltonian systems. Recall that according to Mirowski, neoclassical economists have no legitimate Hamiltonian dynamics because they haven't figured out what is conserved. Further, he says, they have no theoretical rationale for thinking an economic system is dissipative "and hence, no rationale for even looking for strange attractors" (Mirowski 1990, 302). Yet a response to Kelsey by Michael Radzicki notes that dissipative systems need not be defined in terms of loss of energy to friction (or other causes) but merely by a shrinking ("negative divergence") of the flow in state space. In economics, this could be interpreted as depreciation of inventory, death or emigration of population, smoothing of information to formulate expectations, etc. (Radzicki 1988, 692).

14. Conte, for example, seems to associate nonlinear narrative with nonlinear systems (2002, 197).

15. Mirowski is not the only one to express concern for fit in terms of coherence. Alexander Rosenberg identifies a problem of "incoherence" in some attempts to borrow from evolutionary theory to the economic "theory of the firm," because both firms and lineages of firms are identified as the evolutionary interactors (1994, 404). To borrow from mathematical terminology, such a mapping from economics to biology is "many-to-one."

16. Even the folding does not match particularly well, because the process of blending genres is not irreversible. Having constructed a sci-fi opera, for example, the two genres have not become forever combined. But in phase space, trajectories that start in different places can coalesce asymptotically until their separate origins are forever lost to outside observers. Furthermore, in nonlinear dynamics, each phase space point represents one possible configuration of

a given system, so it is difficult to see how such a point could correspond to an entire genre. These particular failures of fit are mismatches between the elements of the metaphor rather than violations of the theoretical coherence of the field as a whole.

17. Adams, Brumwell, and Glazier sometimes seem to suggest that chaos results whenever three or more bodies interact, and may conflate chaotic behavior with the presence of isolated points of instability (1998, 541, 525). But these minor errors of emphasis do not weaken their main point, because they use chaos theory primarily as a metaphor to reorient our thinking about causal relationships. Note also that it would be perfectly possible to challenge the notion of proximate cause without recourse to chaos theory, so the metaphor lacks the virtue of distinctiveness as described later in the chapter.

18. In some cases, a metaphorical borrowing was clearly not necessary and yet would have been apt. Think how helpful it might have been for Milner S. Ball to have the language of chaos theory available when he was making his point about the need for an alternative metaphor for the law based on fluid flow—a metaphor that could flesh out his insight that "order is not truly opposed to chaos" (1985, 138).

19. Thanks to Dave Doyle for first pointing out to me the importance of being aware of potential disanalogies.

CHAPTER 7

1. See David Hume (1739), *A Treatise on Human Nature*, book 3, pt. 1, sec. 1. Some writers, especially those identified as "prescriptivists," hold that all evaluative discourse can be rephrased in terms of "ought" statements. But I use the term "evaluative" because it encompasses a broader scope of ethical, political, and aesthetic concerns. On this point, see Williams (1985, 124–25). Furthermore, the "is–ought" gap and the fact–value dichotomy are not strictly the same issue and do not necessarily imply each other. A sharp distinction between facts and values may mean that talk of values is neither true nor false, or that such talk deals with a different kind of "nonfactual" truth, or that it just deals with different kinds of facts (Dodd and Stern-Gillet 1995, 730–33).

2. Naturalism in ethics traces back to Aristotle, and John Cottingham provides a good explanation of why the word "nature" forms the root of naturalism: the core of naturalism is "the view that an analysis of human nature will yield conclusions about the *requirements for human happiness*" (1983, 462, emphasis in original). A naturalistic ethics, for instance, would use facts about our nature, together with the notion that it is good to live according to our nature, to forge a connection between nonevaluative and evaluative inquiry (Hudson 1969, 29).

3. On the replacement of value discourse by the technical, see Habermas (1988).

4. I have relied here on Jennifer Welchman's (1995) detailed exposition of Dewey's views.

5. A Deweyan notion of human nature would involve reliable empirical generalizations about the way we are, rather than any timeless essence. See Buller

(2005) for a helpful discussion of why humans have no "nature" in the latter sense.

6. Scott notes that he is aware that his assertion places him in danger of being accused of the naturalistic fallacy, but he demurs from addressing this concern.

7. Robert Scott reveals that he places a value on stability when he states that "Chaos Theorists have also come to the conclusion that chaotic (or nonlinear) processes are—because of their unpredictability—more stable than those in equilibrium (linear processes)" (1993, 349). He cites Gleick (1987, 48), but Gleick says no such thing there.

8. See Sen (1987). Regarding these concerns about Pareto optimality, see also Dupré (2001) and Day and Chen (1993, 315–16).

9. For extended critiques of recent work in evolutionary psychology, see Dupré (2001) and Buller (2005).

10. Columnist Dan Savage provides an apt illustration of the way some people conveniently switch between these two strategies: "Homophobes used to argue that homosexuality was unnatural because no other animals engaged in it. When scientists finally admitted that, yes, animals do engage in homosexual acts, the bigots turned around and insisted that homosexuality is disgusting because animals engage in it" (Savage 2004, 101).

11. James Juniper argues that Grandmont's justification for intervention is based on the undesirable quality of chaotic uncertainty but that most economic decision-making is done under conditions of fundamental uncertainty not even related to chaos (2000, 54). So he calls for pluralism in econometric approaches, ultimately offering not a criticism of chaotic models but a rebuke to the exclusive pursuit of any single model.

12. Philip Mirowski rejects claims that chaos theory can show endogenous economic fluctuations that may legitimize governmental intervention (1990, 303). Because he bases this rejection on his view that neoclassical dynamics is fundamentally incoherent, his objection seems to be not to the use of chaos theory but to neoclassical economics itself.

13. Yet Rosser states that one does not need to assume incomplete markets to get endogenous cycles or chaotic behavior in a model (1991, 322). And a recent article states that the conditions needed for chaotic behavior are merely "less stringent in models with incomplete markets, imperfect competition, or externalities" (Guo and Lansing 2002, 657).

14. A similar point is made by Gerald Dwyer (1992), who demonstrates that stabilization policy itself can lead to chaos. From this result he concludes that nonlinearity by itself does not justify intervention.

15. Hayek refers to the socialists' desire to bring the "chaos" of competitive social organization under the control of rational planning (1948, 119, 177). Lavoie calls the calculation argument "essentially an application of spontaneous order analysis to economics" (1989, 627).

16. Mirowski claims that the EMH results from the need to reconcile deterministic theories about equilibrium with the facts that prices fluctuate stochastically (1994b, 472). He further claims that the hypothesis of efficient markets is based on the idea of arbitrage: rational stabilization to prices that reflect fundamental values. But with long dependencies of random length, and sudden

discontinuous jumps (the so-called Joseph and Noah effects described by Benoit Mandelbrot), this arbitrage cannot be accomplished (Mirowski 1990, 297). I leave aside this objection to the EMH, as it rests on discontinuity rather than chaotic dynamics per se.

17. Some persistence in the direction of price changes was found, but nothing that would make it possible to outperform the market (Cunningham 1994, 556).

18. Cunningham says that objections to this hypothesis on the basis of the unrealistic nature of its assumptions were summarily dismissed by invocation of Milton Friedman's philosophy of positive economics, with its injunction that it does not matter how realistic your underlying model is so long as you get good predictions (Cunningham 1994, 559).

19. Barnett and Serletis (2000, 705) identify these definitions as being due to E. F. Fama (1970).

20. The semistrong form of the EMH also depends on the random-walk model being correct, for in this case as well a dependency on past prices means that changes are not entirely due to new information being gained (Cunningham 1994, 561).

21. Nicholas Wolfson reaches a similar conclusion: "Chaos theory requires a radical restructuring of finance theory, law (insofar as it relies upon finance theory), and at least a partial abandonment of the long-held notions of the random walk and efficient market hypothesis" (1989, 516). See also Bausor (1994, 118).

22. Some have attempted to save the EMH by assigning supposed failures to the auxiliary model of capital asset pricing. Such a model is needed in order to specify what is meant by information being "fully reflected" in a price. But Cunningham casts doubts on these pricing models on the basis of their assumption that our preferences are fully rational, and he insists that nonlinear dependencies in security prices are incompatible with the EMH coupled with any pricing model whatsoever (1994, 569–70). Still other auxiliary hypotheses are mentioned in Abhyankar, Copeland, and Wong (1995).

23. Although this intriguing suggestion makes use of a nonlinear perspective—trying to minimize positive feedback of all kinds—it does not seem directly connected to chaos theory.

24. Note that in the final sentence just quoted, Ruhl seems to suggest that the scientific model can itself answer questions about how we ought to act. Indeed, Jeffrey Rudd criticizes Ruhl for attempting to construct a "nonnormative" basis for regulatory reform (Rudd 2005, 554). Although some complexity theorists have advocated this kind of collapse, Ruhl at least sometimes speaks of scientific models informing evaluative choice rather than supplanting it.

25. Economist Giovanni Dosi agrees that "contemporary political experience and urgent normative challenges" should force economists to "think about prescriptions well beyond standard exercises on the Pareto properties of different equilibria" (Day and Chen 1993, 315–16).

26. Similarly, William Demastes asserts that the protagonists' downfall in Ibsen's *The Master Builder* stems from his failure to "embrace uncertainty and to engage in unpredictable vitality" (1998, 84 [quote], 164). See also Seuss (1954).

Daniel Palumbo claims that "the fundamental assumption of chaos theory...is that the individual unit does not matter," so its holistic bent undermines individual agency and autonomy (2002, 33). But this conclusion rests on a drastic misreading of the "holism" in chaos theory. See Kellert (1993, 88–90).

CHAPTER 8

1. William Demastes also speaks of chaos theory as "validating" the suggestions of artists (1998, xiii). Notice how far we are from the more modest claims that science parallels developments in art, or enriches our understanding of it.

2. I do not rehearse here the criticisms of naive sociobiology, which seeks to make ethics a subdiscipline of biology. My discussion owes a debt to Philip Kitcher's (1985) diagnosis of sociobiologists' argumentative strategies. Argyros seeks to escape the charge of naive reductionism by claiming that he does not think "all human behavior is determined by our genetic makeup" (1991, 353). Indeed, he does not simply collapse ethics into biology, nor do some other sociobiologists. For example, see Wilson and Sober (1998) and Hrdy (1981).

3. Argyros even says that "evolution should be viewed as the increasing complexation of time" (1991, 149), as if time itself becomes more complex or forms chemical compounds.

4. The picture of the world as driven toward complexity is associated with the complexity theory of the Santa Fe Institute. For an account of some of the criticisms of this view, see Lewin (1992). Andrew Hayes seems to be inspired by complexity theory more than chaos when he proposes a fractal model of the legal system in which "the natural state of the law is one of constant turbulence," because he speaks of "self-organized criticality" and "non-deterministic" processes (1992, 768).

5. Hawkins describes her work as "careless-and-reckless driving (while intoxicated)" on a "joy-ride through literature in chaos theory's custom-designed Lamborghini" (1995, xii). She also describes herself as "pillaging the writings of chaos theorists" (19) and makes reference to the mysterious, beguiling ways of cats, which "are naturally strange attractors" (127) and whose whiskers make an X, which is the first letter of the Greek word for chaos (167–68). Perhaps it is not appropriate to apply critical scrutiny to a work of such joyous speculation. Yet Hawkins makes pointed arguments against some widely held views. Taking those arguments seriously enough to respond critically should be seen as a sign of respect.

6. Alexander Argyros also confuses universality in Feigenbaum's sense with universality of aesthetic concepts (1991, 281). On the other hand, Thomas Weissert claims that chaos theory argues against global or universal accounts because it shows "radically different behavior among nearby parts of the system" (1991, 233). But the divergence of nearby trajectories does not imply that different parts of the system behave differently.

7. But Indiana goes on to question whether or not the "merely beautiful" is meaningful as art.

8. For fractal analysis applied to music, see Short (1991) as well as Schenker (1973). For fractal analysis applied to the paintings of Jackson Pollack, see

Taylor, Micolich, and Jonas (1999). Additional citations may be found on the Web site for a course on fractal geometry taught by Mandelbrot at Yale: http://classes.yale.edu/fractals/Panorama.

9. Notice the denial of metaphorical translation at work in the use of the word "exactly."

10. Or so I hear.

11. For another discussion of chaotic systems as creating information, see Hayles (1990, 176ff.).

12. The mathematical measurement of information content only works where there is a well-defined ensemble of possible messages. And the founder of information theory, Claude Shannon, deliberately and intentionally circumvented problems of context dependence by explicitly distinguishing "information" from "meaning." Of course, this stringent distinction was no sooner drawn than it was subverted by enthusiastic borrowing facilitated by convenient equivocation. On this point, see Hayles (1990, 270). A further technical problem arises because the class of algorithmically uncomputable systems is not the same as the class of chaotic systems (Leiber 1998, 364).

13. Livingston's scare quotes signal his acknowledgment of the problems with accounts that see science merely constructing images of political arrangements and his desire to avoid collapsing science into the social. For a characterization of fractals as an image of oppositional freedom, see Eglash (1999, 200).

CHAPTER 9

1. Similarly, Claude Richard identifies a central element in the postmodern fiction of Pynchon, Barth, Coover, and others to be the rejection of narrative causality through the "multiplication of unmotivated, unended, incompatible plots and subplots," "coincidences," and "redistribution along aleatory lines of verbal networks" (1983, 86). But such challenges to traditional understandings of causality resonate much more with quantum mechanics than with chaos theory.

2. Conte also explores connections between postmodern novels by authors such as Kathy Acker and Don DeLillo and scientific work on "self-organizing systems" by researchers such as Ilya Prigogine and those at the Santa Fe Institute. As explained in chapter 1, such discussions do not fall within the scope of chaos theory more narrowly defined. On DeLillo, note Gordon Slethaug's assertion that in the novel *White Noise*, death "becomes a strange attractor, generating turbulence and providing pattern" (2000, 148).

3. Note that many literary scholars use the term "Newtonian" as historical shorthand or even as a term of opprobrium. They should not be understood to be referring to Sir Isaac's equations of motion but to the dominant ideological complex of eighteenth-century science. After all, deterministic chaos is perfectly Newtonian in the narrow physical sense of its equations, even though it challenges some central features of traditional scientific methodology (Sokal and Bricmont 1998, 144–45; see also Kellert 1993, chap. 4).

4. "Naturalism" here means something different from what is involved in naturalized epistemology, and different yet again from naturalist ethics.

Demastes locates dramatic naturalism in such authors as Zola, who "borrowed from science the concept of scientific method of observation and experimentation" (1998, 12) and sought to make theater an accurate representation of everyday life.

5. In a similar vein, Conte identifies postmodernist writers as welcoming the interpenetration of order and chaos, "the making of meaning and the free play of signifiers," the continual dance of creation and destruction, as well as dispersion and unpredictability (2002, 9). For instance, he points out that Kathy Acker's *The Empire of the Senseless* (1988) includes the design of a tattoo with a pierced rose and the banner "Discipline and Anarchy." This tattoo provides a powerful image of the linking of freedom and control which also characterizes her method of composition: transgressive and disorderly, yet planned and theoretically informed (Conte 2002, 54).

6. One possible source of confusion stems from the fact that in his enthusiasm for complexity theory, which purports to study the "edge of chaos," Conte repeats the notion popularized by the Santa Fe Institute that complexity is in the middle between the order of periodic behavior and the randomness of the strange attractor (2002, 186). Thus, he fails to note that strange attractors are themselves hybrids exhibiting both order and randomness.

7. Joseph Conte also finds fractal structure in Pynchon's work. He says that the crazy baroque mansion called "The White Visitation" is "a fractal iteration of the design of the whole book in its part" and that "like a Koch curve, *Gravity's Rainbow* constructs an infinite text within a finite space" (Conte 2002, 168, 170). But however dense the allusions in Pynchon's work (and they are damn dense), *Gravity's Rainbow* remains finite at 887 pages long. Conte also allows himself to say that "just as Pynchon's book moves 'Beyond the Zero,' Mandelbrot found it necessary to move beyond integral dimensions of 0, 1, 2, and 3 to a fractional dimensionality in order to describe the irregularity of shapes in nature" (171).

8. A number of other writers have noted the resonances between chaos theory and romanticism, including James Gleick (1987, 163), Ira Livingston (1997, 9–11 and passim), and William Demastes (1998, xiii). Also note the parallels so clearly manifested in Tom Stoppard's *Arcadia* (1993).

9. Moran makes a parallel point regarding the study of law. Postmodern scholars who reject the objectivism and rationality of traditional legal scholarship also find those ideals exemplified in classical science. "It is no surprise, then, that an entire academic movement has been willing to seize hold of these new and astonishing scientific insights as a form of intellectual currency to be wielded at detractors with reversal-of-fortune claim" (Moran 1997, 198).

10. I do not explore connections between postmodern thought and fractals, but recall Ira Livingston's discussion of the self-similar workings of contemporary power as akin to Michel Foucault's analysis of the workings of "discipline" (in chap. 8). And note the enthusiastic discussion of fractals in both Lyotard's *The Postmodern Condition* and Deleuze and Guattari's seminal work *A Thousand Plateaus*. Pursuing these connections would take us a bit too far from the focus on literature, and the connections are not fleshed out in nearly as much detail as in Hayles and Argyros. I also do not spend much time on Jean Baudrillard,

the French postmodern theorist who identifies chaos as reminiscent of the "suppression" of cause-and-effect relationships, and even says it calls to mind Jacques Beneviste's work on the memory of water (1994, 110). He also says that the zero point of an exponentially stable system is a strange attractor (112).

11. Argyros has stated that "the success of deconstruction as an academic paradigm has been nothing short of spectacular" (1991, 1), but critics of an approach sometimes overstate its dominance.

12. Curator Laura Trippi, in her essay for the catalog of an exhibit titled "Strange Attractors," similarly posits that chaos theory and postmodernism are part of a general "cultural rupture" (1989). She goes on to characterize this cultural moment as also including feminism, multiculturalism, "and now even a post-Euclidean geometry."

13. Hayles borrows the mathematical term "isomorphism" to describe the similarity, although an isomorphism implies a much stricter correspondence than she argues for (1990, 184).

14. Harriet Hawkins draws a similar connection between chaos theory and the unpredictable proliferation of meaning described by postmodern, poststructuralist theories such as deconstruction. She claims that chaos theory "gains strong support from poststructuralism" because the many variables involved in producing meaning will never yield a predictable result (1995, 19). According to Hawkins, elements of an artwork can "migrate, mate, and mutate in the minds of individual readers, as well as in successive works of art, in entirely different and unpredictable ways cognate to the dynamics of deterministic chaos and postmodernism alike" (69). This assertion represents one of the very few times when literary theory is claimed to offer support for science, but Hawkins here confuses deterministic chaos with a welter of variables.

15. Thanks to Mike Reynolds for offering this observation. See also Stoicheff (1991, 93).

16. Alexander Argyros agrees with Hayles that deconstruction and chaos theory have different attitudes toward order because their different worldviews lead them to interpret the similar phenomena differently (1991, 238–39).

17. See also Matheson and Kirchhoff (1997, 33).

18. For a discussion of the distinction between chaos and the "forgetting" of initial conditions, see Smith (1991). Gordon Slethaug claims that "the fact that the initial conditions are unpredictable, however, does not in any way alter the deterministic attraction that phenomena take in following certain patterns" (2000, xxv). But I can categorically state that the source he cites (Kellert 1993, xii) never says initial conditions are unpredictable or speaks of "deterministic attraction."

19. Matheson and Kirchhoff criticize Hayles on this same point but choose to focus on a hyperbolic reading of deconstruction: "Where chaos claims that origins exist but are difficult to recover (i.e. to know), Derrida asserts that any origin we may posit is at best a necessary fabrication, a substitute for something that cannot exist" (Matheson and Kirchhoff 1997, 35).

20. On this point, see Kellert (1993, chap. 2).

21. For further confirmation that Derrida is not interested in simply inverting traditional hierarchies, see Armour (1999, 69).

22. Argyros also speaks of deconstruction as holding an "implicit Cartesian dualism" because it separates our conceptual thinking from the knowledge our bodies have as physical objects (1991, 170; see also 176). I do not treat here his notion that rocks have knowledge (168).

23. Even those sympathetic to postmodernism often recur to realist language in saying fractal forms are "in actuality more prevalent" than the smooth shapes of classical geometry (Conte 2002, 10; see also Hayles 1990, 11).

24. Mackey also expresses a somewhat moderate view in saying that "chaos theory provides the fruitful middle ground between the extremes of the post-modern and mechanical models. This theory accepts the indeterminateness and interrelationships emphasized in postmodernism yet affirms the existence of an aboriginal reality" (1999, 19).

25. This is a curious way to label Hayles. After all, she herself is anxious about constructivism and deconstruction because they can remove taboos on destructive behavior and undo useful concepts like human rights (1990, 285). Her view might be better characterized as quasi-Kantian: everything is a representation, with ultimate reality unknowable directly, but some representations are "ruled out by physical constraints" (222–24). Her insistence on the need for such constraints can thus be read as anxiety about relativism.

26. We see here an enactment of what Richard Bernstein calls Cartesian Anxiety: the worry that unless we have a universal, context-free standard, we face an abyss of relativism in which anything goes (1983, 16–19).

27. Ellen Armour has argued convincingly that deconstruction not only avoids relativism but provides useful ethical guidance. She argues that Derrida uncovers the way the concept of "Man" is founded in an unstable relationship to his raced and gendered "Others," and that this very examination issues in a call of political engagement (1999, 154). She also cites Mark C. Taylor, who critiques the claim that deconstruction makes ethical choices impossible in his essay "Not Just Resistance" (1993).

28. Besides the many errors already mentioned, Argyros confuses states with basins of attraction (1991, 242), basins of attraction with attractors (250), and attractors with phase space (252). He also says that frictionless systems have attractors, when they do not (250).

29. See, for example, M. Norton Wise (1993).

Works Cited

Abbott, Andrew. 2001. *Chaos of Disciplines*. Chicago: University of Chicago Press.

Abhyankar, Abhay, Laurence S. Copeland, and W. Wong. 1995. "Nonlinear Dynamics in Real-Time Equity Market Indices: Evidence from the United Kingdom." *Economic Journal* 105:864–80.

———. 1997. "Uncovering Nonlinear Structure in Real-Time Stock-Market Indexes: The S & P 500, the DAX, the Nikkei 225, and the FTSE-100." *Journal of Business and Economic Statistics* 15:1–14.

Acker, Kathy. 1988. *The Empire of the Senseless*. New York: Grove.

Adams, Edward S., Gordon B. Brumwell, and James A. Glazier. 1998. "At the End of Palsgraf, There Is Chaos: An Assessment of Proximate Cause in the Light of Chaos Theory." *University of Pittsburgh Law Review* 507:507–55.

Alborn, Timothy L. 1994. "Economic Man, Economic Machine: Images of Circulation in the Victorian Money Market." In *Natural Images in Economic Thought: "Markets Read in Tooth and Claw,"* edited by Philip Mirowski, 173–96. New York: Cambridge University Press.

Appadurai, Arjun. 1996. *Modernity at Large: Cultural Dimensions of Globalization*. Minneapolis: University of Minnesota Press.

Archard, David, ed. 1996. *Philosophy and Pluralism*. Royal Institute of Philosophy Supplement 40. Cambridge: Cambridge University Press.

Argyros, Alexander J. 1991. *A Blessed Rage for Order: Deconstruction, Evolution, and Chaos*. Ann Arbor: University of Michigan Press.

Armour, Ellen. 1999. *Deconstruction, Feminist Theology, and the Problem of Difference*. Chicago: University of Chicago Press.

Arrigo, Bruce A. 2002. "The Critical Perspective in Psychological Jurisprudence: Theoretical Advances and Epistemological Assumptions." *International Journal of Law and Psychiatry* 25:151–72.

Arthur, Brian W. 1990. "Positive Feedbacks in the Economy." *Scientific American* 262:92–99.

Artigiani, Robert. 1992. "Chaos and Constitutionalism: Toward a Post-Modern Theory of Social Evolution." *World Futures* 34:131–56.

Assad, Maria L. 1991. "Michel Serres: In Search of a Tropography." In *Chaos and Order: Complex Dynamics in Literature and Science*, edited by N. Katherine Hayles, 278–98. Chicago: University of Chicago Press.

Bak, Per, K. Chen, José Scheinkman, and M. Woodford. 1993. "Aggregate Fluctuations from Independent Sectoral Shocks: Self-Organized Criticality in a Model of Production and Inventory Dynamics." *Ricerche Economiche* 47:3–30.

Ball, Milner S. 1985. *Lying Down Together: Law, Metaphor, and Theology*. Madison: University of Wisconsin Press.

Bandes, Susan. 1999. "*Terry v. Ohio* in Hindsight: The Perils of Predicting the Past." *Constitutional Commentary* 16:491–97.

Barnes, Barry, and David Bloor. 1982. "Relativism, Rationalism and the Sociology of Knowledge." In *Rationality and Relativism*, edited by Martin Hollis and Steven Lukes, 21–47. Oxford: Blackwell.

Barnett, William A. 2006. "Comments on 'Chaotic Monetary Dynamics with Confidence.'" *Journal of Macroeconomics* 28:253–55.

Barnett, William A., and Ping Chen. 1988. "The Aggregation–Theoretic Monetary Aggregates Are Chaotic and Have Strange Attractors: An Econometric Application of Mathematical Chaos." In *Dynamic Econometric Modeling*, edited by William A. Barnett, Ernst R. Berndt, and Halbert White, 199–246. Cambridge: Cambridge University Press.

Barnett, William A., and Seungmook S. Choi. 1989. "A Comparison between the Conventional Econometric Approach to Structural Inference and the Nonparametric Chaotic Attractor Approach." In *Economic Complexity: Chaos, Sunspots, Bubbles, and Nonlinearity*, edited by William A. Barnett, John Geweke, and Karl Shell, 141–212. Cambridge: Cambridge University Press.

Barnett, William A., and Melvin J. Hinich. 1993. "Has Chaos Been Discovered with Economic Data?" In *Nonlinear Dynamics and Evolutionary Economics*, edited by Richard H. Day and Ping Chen, 254–65. New York: Oxford University Press.

Barnett, William A., and Apostolos Serletis. 2000. "Martingales, Nonlinearity, and Chaos." *Journal of Economic Dynamics and Control* 24:703–24.

Barondes, Royce de R. 1995. "The Limits of Quantitative Legal Analyses: Chaos in Legal Scholarship and *FDIC v. W. R. Grace & Co.*" *Rutgers Law Review* 48:161–225.

Barth, John. 1995. *Further Fridays: Essays, Lectures, and Other Nonfiction, 1984–94*. Boston: Little, Brown.

Baudrillard, Jean. 1994. *The Illusion of the End*. Translated by Chris Turner. Stanford: Stanford University Press.

Bauer, Henry H. 1990. "Barriers against Interdisciplinarity: Implications for Studies of Science, Technology, and Society (STS)." *Science, Technology, and Human Values* 15:105–19.

Bauman, Richard W. 1996. *Critical Legal Studies: A Guide to the Literature*. Boulder, CO: Westview Press.

Baumol, William J., and Jess Benhabib. 1989. "Chaos: Significance, Mechanism, and Economic Applications." *Journal of Economic Perspectives* 3: 77–105.

Bausor, Randall. 1994. "Qualitative Dynamics in Economics and Fluid Mechanics: A Comparison of Recent Applications." In *Natural Images in Economic Thought: "Markets Read in Tooth and Claw,"* edited by Philip Mirowski, 109–27. New York: Cambridge University Press.

Bechtel, William. 1987. "Psycholinguistics as a Case of Cross-Disciplinary Research: Symposium Introduction." *Synthese* 72:293–311.

Bechtold, B. L. 1997. "Chaos Theory as a Model for Strategy Development." *Empowerment in Organisations* 5:193–201.

Benhabib, Jess, and Richard H. Day. 1982. "A Characterization of Erratic Trajectories in the Overlapping Generations Model." *Journal of Economic Dynamics and Control* 2:37–55.

Bernstein, Richard. 1983. *Beyond Objectivism and Relativism: Science, Hermeneutics, and Praxis*. Philadelphia: University of Pennsylvania Press.

Black, Max. 1962. *Models and Metaphors*. Ithaca, NY: Cornell University Press.

———. 1993. "More about Metaphor." In *Metaphor and Thought*, 2nd ed., edited by Andrew Ortony, 19–41. New York: Cambridge University Press.

Blamires, Harry. 1976. *The Bloomsday Book*. New York: Methuen.

Bloor, D. 1976. *Knowledge and Social Imagery*. London: Routledge and Kegan Paul.

Boldrin, Michele, and Luigi Montrucchio. 1986. "On the Indeterminacy of Capital Accumulation Paths." *Journal of Economic Theory* 40:26–39.

Boyd, Richard. 1993. "Metaphor and Theory Change: What Is 'Metaphor' a Metaphor For?" In *Metaphor and Thought*, 2nd ed., edited by Andrew Ortony, 481–532. New York: Cambridge University Press.

Briggs, John. 1992. *Fractals: The Patterns of Chaos*. London: Thames and Hudson.

Briggs, John, and F. David Peat. 1989. *Turbulent Mirror: An Illustrated Guide to Chaos Theory and the Science of Wholeness*. New York: Harper and Row.

Brion, Denis J. 1995. "The Chaotic Indeterminacy of Tort Law: Between Formalism and Nihilism." In *Radical Philosophy of Law*, edited by David S. Caudill and Steven Jay Gold, 180–99. Atlantic Highlands, NJ: Humanities Press.

Brock, William A., W. D. Dechert, and José Scheinkman. 1986. "A Test for Independence Based on the Correlation Dimension." Manuscript. University of Wisconsin.

Brock, William A., W. D. Dechert, José Scheinkman, and B. LeBaron. 1996. "A Test for Independence Based on the Correlation Dimension." *Econometric Reviews* 15:197–235.

Brock, William A., and Cars H. Hommes. 1997. "A Rational Route to Randomness." *Econometrica* 65:1059–95.

Brock, William A., and A. G. Malliaris. 1989. *Differential Equations, Stability and Chaos in Dynamic Economics.* New York: North-Holland.

Brown, Harold I. 1988. "Normative Epistemology and Naturalized Epistemology." *Inquiry* 31:53–78.

Brown, Richard H. 1977. *A Poetic for Sociology: Toward a Logic of Discovery for the Human Sciences.* Cambridge: Cambridge University Press.

Buchanan, James M., and Viktor J. Vanberg. 1991. "The Market as a Creative Process." *Economics and Philosophy* 7:167–86.

Bullard, James, and Alison Butler. 1993. "Nonlinearity and Chaos in Economic Models: Implications for Policy Decisions." *Economic Journal* 103:849–67.

Buller, David J. 2005. *Adapting Minds: Evolutionary Psychology and the Persistent Quest for Human Nature.* Cambridge: MIT Press.

Butz, Michael R. 1992. "The Fractal Nature of the Development of the Self." *Psychological Reports* 71:1043–63.

Campbell, Donald T. 1969. "Ethnocentrism of Disciplines and the Fish–Scale Model of Omniscience." In *Interdisciplinary Relationships in the Social Sciences,* edited by Muzaher Sherif and Carolyn W. Sherif, 328–48. Chicago: Aldine Publishing.

Cardim de Carvalho, Fernando J. 2003. "Decision-Making under Uncertainty as Drama: Keynesian and Shaklean Themes in Three of Shakespeare's Tragedies." *Journal of Post Keynesian Economics* 25:189–218.

Carrington, Paul D. 1992. "Butterfly Effects: The Possibilities of Law Teaching in a Democracy." *Duke Law Journal* 41:741–805.

Cartwright, Nancy. 1999. *The Dappled World: A Study of the Boundaries of Science.* Cambridge: Cambridge University Press.

Cartwright, T. J. 1991. "Planning and Chaos Theory." *Journal of the American Planning Association* 57:44–56.

Casey, Edward. 1997. *The Fate of Place: A Philosophical History.* Berkeley: University of California Press.

Cass, David, and Karl Shell. 1989. "Sunspot Equilibrium in an Overlapping-Generations Economy with an Idealized Contingent-Commodities Market." In *Economic Complexity: Chaos, Sunspots, Bubbles, and Nonlinearity,* edited by William A. Barnett, John Geweke, and Karl Shell, 3–20. Cambridge: Cambridge University Press.

Cellarier, Laurent. 2006. "Constant Gain Learning and Business Cycles." *Journal of Macroeconomics* 28:51–85.

Chaitin, Gregory. 1977. "Algorithmic Information Theory." *IBM Journal of Research and Development* 21:350–59.

Chen, Jim, ed. 1999. "The Sound of Legal Thunder: The Chaotic Consequences of Crushing Constitutional Butterflies." Special issue. *Constitutional Commentary* 16:483–598.

Chen, Ping. 1993. "Searching for Economic Chaos: A Challenge to Econometric Practice and Nonlinear Tests." In *Nonlinear Dynamics and Evolutionary Economics*, edited by Richard H. Day and Ping Chen, 217–53. New York: Oxford University Press.

Chiarella, Carl, and Peter Flaschel. 2000. "The Emergence of Complex Dynamics in a 'Naturally' Nonlinear Integrated Keynesian Model of Monetary Growth." In *Commerce, Complexity, and Evolution: Topics in Economics, Finance, Marketing and Management*, edited by William Barnett, Carl Chiarella, Steve Keen, Robert Marks, and Herman Schnabl, 111–45. Cambridge: Cambridge University Press.

Christiano, L. J., and S. G. Harrison. 1999. "Chaos, Sunspots and Automatic Stabilizers." *Journal of Monetary Economics* 44:3–31.

Clark, Robert C. 1981. "The Interdisciplinary Study of Legal Evolution." *Yale Law Journal* 90:1238–74.

Cohen, Avi J. 1993. "What Was Abandoned Following the Cambridge Capital Controversies? Samuelson, Substance, Scarcity, and Value." In *Non-Natural Social Science: Reflecting on the Enterprise of More Heat Than Light*, edited by Neil de Marchi, 202–19. Annual supplement to vol. 25 of *History of Political Economy*. Durham, NC: Duke University Press.

Cohen, I. Bernard. 1993. "Analogy, Homology, and Metaphor in the Interactions between the Natural Sciences and the Social Sciences, Especially Economics." In *Non-Natural Social Science: Reflecting on the Enterprise of More Heat Than Light*, edited by Neil de Marchi, 8–44. Annual supplement to vol. 25 of *History of Political Economy*. Durham, NC: Duke University Press.

———. 1994. "Newton and the Social Sciences, with Special Reference to Economics, or, the Case of the Missing Paradigm." In *Natural Images in Economic Thought: "Markets Read in Tooth and Claw,"* edited by Philip Mirowski, 55–90. New York: Cambridge University Press.

Conte, Joseph M. 2002. *Design and Debris: A Chaotics of Postmodern American Fiction*. Tuscaloosa: University of Alabama Press.

Coover, Robert. 1968. *The Universal Baseball Association, Inc., J. Henry Waugh, Prop*. New York: Plume-NAL.

Corrada, Roberto L. 1995. "Justifying a Search for a Unifying Theory of Unconstitutional Conditions." *Denver University Law Review* 72:1011–30.

Cottingham, John. 1983. "Neo-Naturalism and Its Pitfalls." *Philosophy* 58:455–70.

Cunningham, Lawrence A. 1994. "From Random Walks to Chaotic Crashes: The Linear Genealogy of the Efficient Capital Market Hypothesis." *George Washington Law Review* 62:546–608.

Darden, Lindley, and Nancy Maull. 1977. "Interfield Theories." *Philosophy of Science* 44:43–64.

Davidson, Donald. 1984. *Inquiries into Truth and Interpretation*. Oxford: Clarendon Press.

Day, Richard H. 1993. "Nonlinear Dynamics and Evolutionary Economics."
In *Nonlinear Dynamics and Evolutionary Economics*, edited by Richard H.
Day and Ping Chen, 18–41. New York: Oxford University Press.

Day, Richard H., and Ping Chen. 1993. "Round Table Discussion." Partici-
pants: Peter Allen, Brian Arthur, William Brock, Ping Chen, John Conlisk,
Richard Day, Giovanni Dosi, Sidney Winter, Richard Goodwin, Jacques
Lesourne, Richard Lipsey, Michael Mackey, Richard Nelson, Ilya Prigogine,
Peter Schwefel, and John Sterman. In *Nonlinear Dynamics and Evolutionary
Economics*, edited by Richard H. Day and Ping Chen, 309–24. New York:
Oxford University Press.

de Man, Paul. 1979. *Allegories of Reading: Figural Language in Rousseau, Nietzsche, Rilke, and Proust*. New Haven: Yale University Press.

de Marchi, Neil. 1993. Introduction to *Non-Natural Social Science: Reflecting
on the Enterprise of More Heat Than Light,* edited by Neil de Marchi, 1–4.
Annual supplement to vol. 25 of *History of Political Economy.* Durham,
NC: Duke University Press.

Dechert, W. D., and Cars H. Hommes. 2000. "Complex Nonlinear Dynamics
and Computational Methods." *Journal of Economic Dynamics and Control*
24:651–62.

Demastes, William W. 1998. *Theatre of Chaos: Beyond Absurdism, into Orderly Disorder.* New York: Cambridge University Press.

Derrida, Jacques. 1976. *Of Grammatology.* Translated by Gayatri Chakravorty
Spivak. Baltimore: Johns Hopkins University Press.

———. 1982. *Margins of Philosophy.* Translated by Alan Bass. Chicago: University of Chicago Press.

Dewey, John. [1915] 1979. "The Logic of Judgments of Practice." In *German
Philosophy and Politics and Schools of Tomorrow,* vol. 8 of *John Dewey:
The Middle Works, 1899–1924,* edited by Jo Ann Boydston, 14–82. Carbon-
dale: Southern Illinois University Press.

———. 1934. *Art as Experience.* New York: Perigee.

Dewey, John, and James H. Tufts. [1908] 1978. *Ethics,* vol. 5 of *John Dewey:
The Middle Works, 1899–1924,* edited by Jo Ann Boydston. Carbondale:
Southern Illinois University Press.

DiLorenzo, Vincent M. 1994. "Legislative Chaos: An Exploratory Study." *Yale
Law and Policy Review* 12:425–85.

———. 2000. "Equal Economic Opportunity: Corporate Social Responsibility
in the New Millennium." *University of Colorado Law Review* 71:52–112.

Ditto, W. L., S. N. Rauseo, and M. L. Spano. 1990. "Experimental Control of
Chaos." *Physical Review Letters* 65:3211–14.

Dodd, Julian, and Suzanne Stern-Gillet. 1995. "The Is/Ought Gap, the
Fact/Value Distinction and the Naturalistic Fallacy." *Dialogue* 34:727–45.

Downes, Stephen M. 1993. "Socializing Naturalized Philosophy of Science."
Philosophy of Science 60:452–68.

Downs, Robert C. 1995. "Law and Economics: Nexus of Science and Belief."
Pacific Law Journal 27:1–36.

Dupré, John. 1993. *The Disorder of Things: Metaphysical Foundations of the
Disunity of Science.* Cambridge: Harvard University Press.

———. 2001. *Human Nature and the Limits of Science*. Oxford: Oxford University Press.

Dwyer, Gerald. 1992. "Stabilization Policy Can Lead to Chaos." *Economic Inquiry* 30:40–46.

Dyke, Charles. 1990. "Strange Attraction, Curious Liaison: Clio Meets Chaos." *Philosophical Forum* 21:369–92.

Earman, John. 1989. *World Enough and Space-Time: Absolute versus Relational Theories of Space and Time*. Cambridge: MIT Press.

Eglash, Ron. 1999. *African Fractals: Modern Computing and Indigenous Design*. New Brunswick, NJ: Rutgers University Press.

Ekeland, Ivar. 1988. *Mathematics and the Unexpected*. Chicago: University of Chicago Press.

Ellis, Richard S. 1997. "The Book of Leviticus and the Fractal Geometry of Torah." *Conservative Judaism* 50:27–34.

Eoyang, Eugene. 1989. "Chaos Misread: Or, There's a Wonton in My Soup!" *Comparative Literature Studies* 26:271–84.

Ernest, John. 2006. "Representing Chaos: William Craft's *Running a Thousand Miles for Freedom*." *PMLA: Publications of the Modern Language Association of America* 121:469–83.

Fahnestock, Jeanne. 1999. *Rhetorical Figures in Science*. New York: Oxford University Press.

Fama, E. F. 1970. "Efficient Capital Markets: A Review of Theory and Empirical Work." *Journal of Finance* 25:383–417.

Fine, Arthur. 1996. "Science Made Up: Constructivist Sociology of Scientific Knowledge." In *The Disunity of Science: Boundaries, Contexts, and Power*, edited by Peter Galison and David J. Stump, 231–54. Stanford: Stanford University Press.

Fish, Stanley. 1994. *There's No Such Thing as Free Speech: And It's a Good Thing, Too*. New York: Oxford University Press.

———. 1995. *Professional Correctness: Literary Studies and Political Change*. New York: Oxford University Press.

Flores, Robert. 2002. "A Portrait of Don Quixote from the Palette of Chaos Theory." *Cervantes: Bulletin of the Cervantes Society of America* 22:43–70.

Foley, Duncan K. 1989. "Endogenous Financial-Production Cycles in a Macroeconomic Model." In *Economic Complexity: Chaos, Sunspots, Bubbles, and Nonlinearity*, edited by William A. Barnett, John Geweke, and Karl Shell, 89–99. Cambridge: Cambridge University Press.

Forman, Paul. 1971. "Weimar Culture, Causality, and Quantum Theory, 1918–1927." *Historical Studies in the Physical Sciences* 3:1–115.

Foucault, Michel. 1979. *Discipline and Punish*. New York: Vintage.

Franklin, Allen. 1993. *The Rise and Fall of the Fifth Force: Discovery, Pursuit, and Justification in Modern Physics*. New York: American Institute of Physics.

Frayn, Michael. 1998. *Copenhagen*. New York: Samuel French.

Friedman, Michael. 1997. "Philosophical Naturalism." In *Proceedings and Addresses of the American Philosophical Association*, edited by Eric Hoffman, 71:7–21.

Fuller, Steve. 1996. "Talking Metaphysical Turkey about Epistemological Chicken, and the Poop on Pidgins." In *The Disunity of Science: Boundaries, Contexts, and Power*, edited by Peter Galison and David J. Stump, 170–86. Stanford: Stanford University Press.

Galison, Peter. 1996. "Computer Simulations and the Trading Zone." In *The Disunity of Science: Boundaries, Contexts, and Power*, edited by Peter Galison and David J. Stump, 118–57. Stanford: Stanford University Press.

Gaonkar, Dilip Parameshwar. 1997. "The Idea of Rhetoric in the Rhetoric of Science." In *Rhetorical Hermeneutics: Invention and Interpretation in the Age of Science*, edited by Alan G. Gross and William M. Keith, 25–85. Albany: SUNY Press.

Gentner, Dedre, and Michael Jeziorski. 1993. "The Shift from Metaphor to Analogy in Western Science." In *Metaphor and Thought*, 2nd ed., edited by Andrew Ortony, 447–80. New York: Cambridge University Press.

Geu, Thomas Earl. 1994. "The Tao of Jurisprudence: Chaos, Brain Science, Synchronicity, and the Law." *Tennessee Law Review* 61:933–90.

Giere, Ronald N. 1973. "History and Philosophy of Science: Intimate Relationship or Marriage of Convenience?" *British Journal of Philosophy of Science* 24:282–97.

———. 1988. *Explaining Science: A Cognitive Approach*. Chicago: University of Chicago Press.

———. 1999. *Science without Laws*. Chicago: University of Chicago Press.

Gieryn, Thomas F. 1999. *Cultural Boundaries of Science: Credibility on the Line*. Chicago: University of Chicago Press.

Gleick, James. 1987. *Chaos: Making a New Science*. New York: Viking.

Goeree, J. K., and Cars H. Hommes. 2000. "Heterogeneous Beliefs and the Non-Linear Cobweb Model." *Journal of Economic Dynamics and Control* 24:761–98.

Goldberger, A. 1991. "Is the Normal Heartbeat Chaotic or Homeostatic?" *News in Physiological Sciences* 6:87–91.

Goldstein, Rebecca. 1993. "Strange Attractors." In *Strange Attractors*, 243–76. New York: Viking.

Goodman, Nelson. 1968. *Languages of Art*. Indianapolis: Bobbs-Merrill.

Goodwin, Richard M. 1993a. "A Marx-Keynes-Schumpter Model of Economic Growth and Fluctuation." In *Nonlinear Dynamics and Evolutionary Economics*, edited by Richard H. Day and Ping Chen, 45–57. New York: Oxford University Press.

———. 1993b. "My Erratic Progress toward Economic Dynamics: Remarks Made at Banquet, Tuesday, April 18, 1989." In *Nonlinear Dynamics and Evolutionary Economics*, edited by Richard H. Day and Ping Chen, 303–6. New York: Oxford University Press.

Gordon, Scott. 1989. "Darwin and Political Economy: The Connection Reconsidered." *Journal of the History of Biology* 22:437–59.

Gould, Stephen Jay. 1994. "The Evolution of Life on Earth." *Scientific American* 271:84–91.

Grandmont, Jean-Michel. 1985. "On Endogenous Competitive Business Cycles." *Econometrica* 53:995–1045.

———. 1989. "Local Bifurcations and Stationary Sunspots." In *Economic Complexity: Chaos, Sunspots, Bubbles, and Nonlinearity*, edited by William A. Barnett, John Geweke, and Karl Shell, 45–60. Cambridge: Cambridge University Press.

Grattan-Guinness, Ivor. 1994. "From Virtual Velocities to Economic Action: The Very Slow Arrivals of Linear Programming and Locational Equilibrium." In *Natural Images in Economic Thought: "Markets Read in Tooth and Claw,"* edited by Philip Mirowski, 91–108. New York: Cambridge University Press.

Greeley, Lillian. 1995. "Complexity in the Attention System of the Cognitive Generative Learning Process." In *Chaos and Society*, edited by A. Albert, 371–86. Amsterdam: IOS Press.

Grey, Thomas C. 1983. "Langdell's Orthodoxy." *University of Pittsburgh Law Review* 45:1–53.

Gross, Alan G. 1990. *The Rhetoric of Science.* Cambridge: Harvard University Press.

Gross, Paul R., and Norman Levitt. 1994. *Higher Superstition: The Academic Left and Its Quarrels with Science.* Baltimore: Johns Hopkins University Press.

Gumerman, George J., and David A. Phillips Jr. 1978. "Archaeology beyond Anthropology." *American Antiquity* 43:184–91.

Gunter, H. 1995. "Jurassic Management: Chaos and Management Development in Educational Institutions." *Journal of Educational Administration* 33:5–20.

Guo, Jang-Ting, and Kevin J. Lansing. 2002. "Fiscal Policy, Increasing Returns, and Endogenous Fluctuations." *Macroeconomic Dynamics* 6:633–64.

Haack, Susan. 1998. *Manifesto of a Passionate Moderate: Unfashionable Essays.* Chicago: University of Chicago Press.

Habermas, Jürgen. 1988. *Theory and Practice.* Translated by John Viertel. Boston: Beacon Press.

Hacking, Ian. 1999. *The Social Construction of What?* Cambridge: Harvard University Press.

Hahn, Hans. 1954. "Geometry and Intuition." *Scientific American* 190: 84–91.

Haigh, C. 2002. "Using Chaos Theory: The Implications for Nursing." *Journal of Advanced Nursing* 37:462–69.

Hands, D. Wade. 1993. "More Light on Integrability, Symmetry, and Utility as Potential Energy in Mirowski's Critical History." In *Non-Natural Social Science: Reflecting on the Enterprise of More Heat Than Light*, edited by Neil de Marchi, 118–30. Annual supplement to vol. 25 of *History of Political Economy*. Durham, NC: Duke University Press.

Hansson, Bengt. 1999. "Interdisciplinarity: For What Purpose?" *Policy Sciences* 32:339–43.

Hardwig, John. 1985. "Epistemic Dependence." *Journal of Philosophy* 82: 335–49.

Harris Poll. 2004. "Doctors, Scientists, Firemen, Teachers, and Military Officers Top List as 'Most Prestigious Occupations.'" Poll no. 65.

http://www.harrisinteractive.com/harris_poll/index.asp?pid=494E (accessed January 16, 2006).

Hawkes, John. 1976. *Travesty*. New York: New Directions.

Hawkins, Harriet. 1995. *Strange Attractors: Literature, Culture and Chaos Theory*. New York: Prentice Hall.

Hayek, Friedrich A. 1948. *Individualism and Economic Order*. Chicago: University of Chicago Press.

———. 1952. *The Counter-Revolution of Science: Studies on the Abuse of Reason*. Glencoe, IL: Free Press.

Hayes, Andrew W. 1992. "An Introduction to Chaos and Law." *UMKC Law Review* 60:751–73.

Hayles, N. Katherine. 1984. *The Cosmic Web: Scientific Field Models and Literary Strategies in the Twentieth Century*. Ithaca, NY: Cornell University Press.

———. 1990. *Chaos Bound: Orderly Disorder in Contemporary Literature and Science*. Ithaca, NY: Cornell University Press.

———, ed. 1991. *Chaos and Order: Complex Dynamics in Literature and Science*. Chicago: University of Chicago Press.

Hegel, G. W. F. [1830] 1975. *Logic (Part I of the Encyclopedia of the Philosophical Sciences)*. Translated by William Wallace. Oxford: Oxford University Press.

Hentschel, Klaus. 1990. "Philosophical Interpretations of Relativity Theory: 1910–1930." In *PSA 1990: Proceedings of the Biennial Meeting of the Philosophy of Science Association*, 2:169–179.

Hesse, Mary B. 1966. *Models and Analogies in Science*. Notre Dame, IN: University of Notre Dame Press.

Hinich, Melvin J., and Douglas M. Patterson. 1993. "Intraday Nonlinear Behavior of Stock Prices." In *Nonlinear Dynamics and Evolutionary Economics*, edited by Richard H. Day and Ping Chen, 201–14. New York: Oxford University Press.

Hodgson, Geoffrey M. 1994. "Hayek, Evolution, and Spontaneous Order." In *Natural Images in Economic Thought: "Markets Read in Tooth and Claw,"* edited by Philip Mirowski, 408–47. New York: Cambridge University Press.

Holling, C. S. 1973. "Resilience and Stability of Ecological Systems." *Annual Review of Ecology and Systematics* 4:1–24.

Howells, Louise A. 1997. "Looking for the Butterfly Effect: An Analysis of Urban Economic Development under the Community Development Block Grant Program." *Saint Louis University Public Law Review* 16:383–417.

Hoyningen-Huene, Paul. 1987. "Context of Discovery and Context of Justification." *Studies in the History and Philosophy of Science* 18:501–15.

Hrdy, Sarah Blaffer. 1981. *The Woman That Never Evolved*. Cambridge: Harvard University Press.

Huang, Weihong, and Richard H. Day. 1993. "Chaotically Switching Bear and Bull Markets: The Derivation of Stock Price Distributions from Behavioral Rules." In *Nonlinear Dynamics and Evolutionary Economics*, edited by Richard H. Day and Ping Chen, 169–182. New York: Oxford University Press, 1993.

Hudson, Christopher. 2000. "At the Edge of Chaos: A New Paradigm for Social Work?" *Journal of Social Work Education* 36:215–30.

Hudson, W. D. 1969. "Editor's Introduction: The 'Is–Ought' Problem." In *The Is–Ought Question,* edited by W. D. Hudson, 11–31. London: Macmillan.

Hughes, Linda K., and Michael Lund. 1991. "Linear Stories and Circular Visions: The Decline of the Victorian Serial." In *Chaos and Order: Complex Dynamics in Literature and Science,* edited by N. Katherine Hayles, 167–94. Chicago: University of Chicago Press.

Hull, David L. 1988. *Science as a Process: An Evolutionary Account of the Social and Conceptual Development of Science.* Chicago: University of Chicago Press.

Iannone, Ron. 1995. "Chaos Theory and Its Implications for Curriculum and Teaching." *Education* 115:541–46.

Imwinkelried, Edward J. 1995. "Evidence Law Visits Jurassic Park: The Far-Reaching Implication of the *Daubert* Court's Recognition of the Uncertainty of the Scientific Enterprise." *Iowa Law Review* 81:55–78.

Indiana, Gary. 1986. "Chaos Plus." *Village Voice,* October 7.

Jensen, Richard V., and Robin Urban. 1984. "Chaotic Behavior in a Non-Linear Cobweb Model." *Economics Letters* 15:235–40.

Johnston, Jason Scott. 1991. "Uncertainty, Chaos, and the Torts Process: An Economic Analysis of Legal Form." *Cornell Law Review* 76:341–400.

Jones, David, and John Culliney. 1999. "The Fractal Self and the Organization of Nature: The Daoist Sage and Chaos Theory." *Zygon* 34:643–54.

Joyce, James. [1922]1986. *Ulysses.* Edited by Hans Walter Gabler. New York: Vintage.

Juniper, James. 2000. "Uncertainty, Risk, and Chaos." In *Commerce, Complexity, and Evolution: Topics in Economics, Finance, Marketing and Management,* edited by William Barnett, Carl Chiarella, Steve Keen, Robert Marks, and Herman Schnabl, 37–60. Cambridge: Cambridge University Press.

Kaas, Leo. 1998. "Stabilizing Chaos in a Dynamic Macroeconomic Model." *Journal of Economic Behavior and Organization* 33:313–32.

Kant, Immanuel. [1787] 1929. *Critique of Pure Reason.* Translated by Norman Kemp Smith. New York: St. Martin's Press.

Kehr, Dave. 2004. "A Man with a Past Best Forgotten Goes to All Lengths to Remember." *New York Times,* January 23, Arts section.

Keller, Evelyn Fox. 2002. *Making Sense of Life: Explaining Biological Development with Models, Metaphors, and Machines.* Cambridge: Harvard University Press.

Kellert, Stephen H. 1992. "A Philosophical Evaluation of the Chaos Theory 'Revolution.'" In *Proceedings of the Philosophy of Science Association,* edited by D. Hull, M. Forbes, and K. Okruhlik, 33–49.

———. 1993. *In the Wake of Chaos: Unpredictable Order in Dynamical Systems.* Chicago: University of Chicago Press.

———. 1995. "When Is the Economy Not Like the Weather? The Problem of Extending Chaos Theory to the Social Sciences." In *Chaos and Society,* edited by A. Albert, 35–47. Amsterdam: IOS Press.

———. 1996. "Science and Literature and Philosophy: The Case of Chaos Theory and Deconstruction." *Configurations* 4:215–32.

———. 2001. "Extrascientific Uses of Physics: The Case of Nonlinear Dynamics and Legal Theory." *Philosophy of Science* 68 (proceedings): S455–S466.

———. 2005. "The Uses of Borrowed Knowledge: Chaos Theory and Antidepressants." *Philosophy, Psychiatry and Psychology* 12:239–42.

———. 2006. "Disciplinary Pluralism for Science Studies." In *Scientific Pluralism: Minnesota Studies in the Philosophy of Science*, vol. 19, edited by Stephen H. Kellert, Helen E. Longino, and C. Kenneth Waters, 215–30. Minneapolis: University of Minnesota Press.

Kellert, Stephen H., Helen E. Longino, and C. Kenneth Waters, eds. 2006. *Scientific Pluralism*. Minneapolis: University of Minnesota Press.

Kelman, Mark. 1987. *A Guide to Critical Legal Studies*. Cambridge: Harvard University Press.

Kelsey, David. 1988. "The Economics of Chaos or the Chaos of Economics." *Oxford Economic Papers* 40:1–31.

Kim, Jaegwon. 1994. "What Is 'Naturalized Epistemology?'" In *Naturalizing Epistemology*, 2nd ed., edited by Hilary Kornblith, 33–56. Cambridge: MIT Press.

Kirby, Michael. 1990. "Chaos and the Rational Mind." *Australian Journal of Forensic Sciences* 22:62–67.

Kitcher, Phillip. 1985. *Vaulting Ambition: Sociobiology and the Quest for Human Nature*. Cambridge: MIT Press.

———. 1990. "The Division of Cognitive Labor." *Journal of Philosophy* 87: 5–22.

———. 1992. "The Naturalists Return." *Philosophical Review* 101:53–114.

———. 1998. "A Plea for Science Studies." In *A House Built on Sand: Exposing Postmodernist Myths about Science*, edited by Noretta Keortge, 32–56. New York: Oxford University Press.

Kittay, Eva Feder. 1987. *Metaphor: Its Cognitive Force and Linguistic Structure*. New York: Oxford University Press.

Klamer, Arjo, and Thomas C. Leonard. 1994. "So What's an Economic Metaphor?" In *Natural Images in Economic Thought: "Markets Read in Tooth and Claw,"* edited by Philip Mirowski, 20–51. New York: Cambridge University Press.

Klein, Julie Thompson. 1990. *Interdisciplinarity: History, Theory, and Practice*. Detroit: Wayne State University Press.

Knoespel, Kenneth J. 1991. "The Emplotment of Chaos: Instability and Narrative Order." In *Chaos and Order: Complex Dynamics in Literature and Science*, edited by N. Katherine Hayles, 100–122. Chicago: University of Chicago Press.

Kockelmans, Joseph J. 1979. "Why Interdisciplinarity?" In *Interdisciplinarity and Higher Education*, edited by Joseph J. Kockelmans, 123–60. University Park: Pennsylvania State University Press.

Kornblith, Hilary. 1994. "Beyond Foundationalism and the Coherence Theory." In *Naturalizing Epistemology*, 2nd ed., edited by Hilary Kornblith, 131–46. Cambridge: MIT Press.

Kuhn, Thomas S. 1970. *The Structure of Scientific Revolutions*. 2nd ed. Chicago: University of Chicago Press.

———. 1977. *The Essential Tension: Selected Studies in Scientific Tradition and Change*. Chicago: University of Chicago Press.

———. 1993. "Metaphor in Science." In *Metaphor and Thought*, 2nd ed., edited by Andrew Ortony, 533–42. New York: Cambridge University Press.

Kundert-Gibbs, John Leelan. 1999. *No-thing Is Left to Tell: Zen/Chaos in the Dramatic Art of Samuel Beckett*. Madison, WI: Farleigh Dickinson University Press.

Lagos, Ricardo, and Randall Wright. 2003. "Dynamics, Cycles, and Sunspot Equilibria in 'Genuinely Dynamic, Fundamentally Disaggregative' Models of Money." *Journal of Economic Theory* 109:156–71.

Lagueux, Maurice. 1999. "Do Metaphors Affect Economic Theory?" *Economics and Philosophy* 15:1–22.

Lakoff, George, and Mark Johnson. 1980. *Metaphors We Live By*. Chicago: University of Chicago Press.

Langer, Susanne K. 1957. *Philosophy in a New Key: A Study in the Symbolism of Reason, Rite, and Art*. 3rd ed. Cambridge: Harvard University Press.

Latour, Bruno. 1987. *Science in Action*. Cambridge: Harvard University Press.

Laudan, Larry. 1987. "Progress or Rationality? The Prospects for Normative Naturalism." *American Philosophical Quarterly* 24:19–31.

———. 1990. "Normative Naturalism." *Philosophy of Science* 57:44–59.

Lavoie, Don. 1989. "Economic Chaos or Spontaneous Order?" *Cato Journal* 8:613–35.

Leff, Arthur Allen. 1974. "Economic Analysis of Law: Some Realism about Nominalism." *Virginia Law Review* 60:451–76.

Leiber, Theodor. 1998. "On the Actual Impact of Deterministic Chaos." *Synthese* 113:357–79.

Lemieux, Pierre. 1994. "Chaos, Complexity, and Anarchy." *Liberty* 7:21–29.

Levit, Nancy. 1996. "Defining Cutting Edge Scholarship: Feminism and Criteria of Rationality." *Chicago-Kent Law Review* 71:947–68.

Lewin, Roger. 1992. *Complexity: Life at the Edge of Chaos*. New York: Macmillan.

Lighthill, James. 1986. "The Recently Recognized Failure of Predictability in Newtonian Dynamics." *Proceedings of the Royal Society of London A* 407:35–50.

Lindenfeld, David F. 1999. "Causality, Chaos Theory, and the End of the Weimar Republic: A Commentary on Henry Turner's *Hitler's Thirty Days to Power*." *History and Theory* 38:281–306.

Limoges, Camille, and Claude Ménard. 1994. "Organization and the Division of Labor: Biological Metaphors at Work in Alfred Marshall's *Principles of Economics*." In *Natural Images in Economic Thought: "Markets Read in Tooth and Claw*," edited by Philip Mirowski, 336–59. New York: Cambridge University Press.

Livingston, Ira. 1997. *Arrow of Chaos: Romanticism and Postmodernity*. Minneapolis: University of Minnesota Press.

Lonçã, Francisco. 2000. "Complexity, Chaos, or Randomness: Ragnar Frisch and the Enigma of the Lost Manuscript." In *Complexity and the History of Economic Thought*, edited by D. Colander, 116–25. New York: Routledge.

Longino, Helen E. 1990. *Science as Social Knowledge: Values and Objectivity in Scientific Inquiry*. Princeton: Princeton University Press.

———. 1993. "Subjects, Power, and Knowledge: Description and Prescription in Feminist Philosophies of Science." In *Feminist Epistemologies*, edited by Linda Alcoff and Elizabeth Potter, 101–20. New York: Routledge.

———. 2002. *The Fate of Knowledge*. Princeton: Princeton University Press.

———. 2006. "Theoretical Pluralism and the Scientific Study of Behavior." In *Scientific Pluralism*, edited by Stephen H. Kellert, C. Kenneth Waters, and Helen E. Longino, 102–31. Minneapolis: University of Minnesota Press.

Lucas, R. E., Jr. 1975. "An Equilibrium Model of the Business Cycle." *Journal of Political Economy* 83:1113–44.

Lyne, John. 1990. "Bio-Rhetorics: Moralizing the Life Sciences." In *The Rhetorical Turn: Invention and Persuasion in the Conduct of Inquiry*, edited by Herbert W. Simons, 35–57. Chicago: University of Chicago Press.

Lyotard, Jean-François. 1984. *The Postmodern Condition: A Report on Knowledge*. Translated by Geoff Bennington and Brian Massumi. Minneapolis: University of Minnesota Press.

Mackey, Peter Francis. 1999. *Chaos Theory and James Joyce's Everyman*. Gainesville: University Press of Florida.

Mandelbrot, Benoit B., and Richard L. Hudson. 2004. *The (Mis)Behavior of Markets: A Fractal View of Risk, Ruin, and Reward*. New York: Basic Books.

Mano, D. K. 1989. "The Love of Randomness." *National Review*, October 13.

Markley, Robert. 1991. "Representing Order: Natural Philosophy, Mathematics, and Theology in the Newtonian Revolution." In *Chaos and Order: Complex Dynamics in Literature and Science*, edited by N. Katherine Hayles, 125–48. Chicago: University of Chicago Press.

Martin-Vallas, Francois. 2005. "Towards a Theory of Integration of the Other in Representation." *Journal of Analytical Psychology* 30:285–93.

Matheson, Carl, and Evan Kirchhoff. 1997. "Chaos and Literature." *Philosophy and Literature* 21:28–45.

Matilla-Garcia, Mariano. 2007. "Nonlinear Dynamics in Energy Futures." *Energy Journal* 28:7–30.

McCarthy, John A. 2006. *Remapping Reality: Chaos and Complexity in Science and Literature*. Amsterdam: Rodopi.

McCloskey, D. N. 1985. *The Rhetoric of Economics*. Madison: University of Wisconsin Press.

———. 1991. "History, Differential Equations, and the Problem of Narration." *History and Theory* 30:21–36.

McManus, Theodore F. [1915] 1959. "The Penalty of Leadership." *Saturday Evening Post*, January 2. Reprinted in Julian Lewis Watkins, *The 100 Greatest Advertisements*. New York: Dover.

Miele, Frank. 2000. "Special Section Introduction: A Quick and Dirty Guide to Chaos and Complexity Theory—Three Race Horses and Four Hobby Horses." *Skeptic* 8:54–60.

Mill, John Stuart. 1886. *System of Logic: Ratiocinative and Inductive.* London: Longmans, Green.

Milovanovic, Dragan. 1997a. Introduction to *Chaos, Criminology, and Social Justice: The New Orderly (Dis)Order,* edited by Dragan Milovanovic, vii–xiv. Westport, CT: Praeger.

———. 1997b. "Visions of the Emerging Orderly (Dis)Order." In *Chaos, Criminology, and Social Justice: The New Orderly (Dis)Order,* edited by Dragan Milovanovic, 194–211. Westport, CT: Praeger.

Mirowski, Philip. 1989. *More Heat Than Light: Economics as Social Physics, Physics as Nature's Economics.* Cambridge: Cambridge University Press.

———. 1990. "From Mandelbrot to Chaos in Economic Theory." *Southern Economic Journal* 57:289–307.

———. 1994a. "Doing What Comes Naturally: Four Metanarratives on What Metaphors Are For." In *Natural Images in Economic Thought: "Markets Read in Tooth and Claw,"* edited by Philip Mirowski, 3–19. New York: Cambridge University Press.

———. 1994b. "The Realms of the Natural." In *Natural Images in Economic Thought: "Markets Read in Tooth and Claw,"* edited by Philip Mirowski, 451–83. New York: Cambridge University Press.

Moore, G. E. 1951. *Principia Ethica.* Cambridge: Cambridge University Press.

Moran, Jay P. 1997. "Postmodernism's Misguided Place in Legal Scholarship: Chaos Theory, Deconstruction, and Some Insights from Thomas Pynchon's Fiction." *Southern California Interdisciplinary Law Journal* 6:155–99.

Moran, Joe. 2002. *Interdisciplinarity.* London: Routledge.

Morin, Edgar. 1983. "Beyond Determinism: The Dialogue of Order and Disorder." Translated by Frank Coppay. *SubStance* 40:22–35.

Morris, Adelaide. 1991. "Science and the Mythopoeic Mind: The Case of H.D." In *Chaos and Order: Complex Dynamics in Literature and Science,* edited by N. Katherine Hayles, 195–220. Chicago: University of Chicago Press.

Mosekilde, Erik, Erik Reimer Larsen, John D. Sterman, and Jesper Skovhus Thomsen. 1993. "Mode Locking and Nonlinear Entrainment of Macroeconomic Cycles." In *Nonlinear Dynamics and Evolutionary Economics,* edited by Richard H. Day and Ping Chen, 58–83. New York: Oxford University Press.

Mukherji, Anjan. 1999. "A Simple Example of Complex Dynamics." *Economic Theory* 14:741–49.

Murphy, James Bernard. 1994. "The Kinds of Order in Society." In *Natural Images in Economic Thought: "Markets Read in Tooth and Claw,"* edited by Philip Mirowski, 536–82. New York: Cambridge University Press.

Naftulin, Donald H., J. E. Ware Jr., and F. A. Donnelly. 1973. "The Doctor Fox Lecture: A Paradigm of Educational Seduction." *Journal of Medical Education* 48:630–35.

Nelson, John S. 1990. "Political Foundations for the Rhetoric of Inquiry." In *The Rhetorical Turn: Invention and Persuasion in the Conduct of Inquiry,* edited by Herbert W. Simons, 258–89. Chicago: University of Chicago Press.

New Museum of Contemporary Art. 1989. *Strange Attractors: Signs of Chaos.* Catalogue organized by Alice Yang. New York: New Museum of Contemporary Art.

Nickles, Thomas. 1980. "Introductory Essay: Scientific Discovery and the Future of Philosophy of Science." In *Scientific Discovery, Logic, and Rationality*, edited by Thomas Nickles, 1–59. Dordrecht, Netherlands: D. Reidel.

———. 1998. "Logic of Discovery." In *The Routledge Encyclopedia of Philosophy*, edited by Edward Craig, 3:99–103. London: Routledge.

———. 2000. "Discovery." In *A Companion to the Philosophy of Science*, edited by W. H. Newton-Smith, 85–96. Oxford: Blackwell.

Nicola, PierCarlo. 2000. *Mainstream Mathematical Economics in the Twentieth Century*. New York: Springer.

Niman, Neil B. 1994. "The Role of Biological Analogies in the Theory of the Firm." In *Natural Images in Economic Thought: "Markets Read in Tooth and Claw,"* edited by Philip Mirowski, 360–83. New York: Cambridge University Press.

Nishimura, Kazuo, and Makoto Yano. 1995. "Nonlinear Dynamics and Chaos in Optimal Growth: An Example." *Econometrica* 63:981–1001.

Nissani, Moti. 1995. "Fruits, Salads, and Smoothies: A Working Definition of Interdisciplinarity." *Journal of Educational Thought* 29:119–26.

———. 1997. "Ten Cheers for Interdisciplinarity: The Case for Interdisciplinary Knowledge and Research." *Social Science Journal* 34:201–16.

Nozick, Robert. 1981. *Philosophical Explanations*. Cambridge: Harvard University Press.

Oliver, Kelly. 1991. "Fractal Politics: How to Use the Subject." *Praxis International* 11:178–94.

Onozaki, Tamotsu, Gernot Sieg, and Masanori Yokoo. 2003. "Affiliation Stability, Chaos and Multiple Attractors: A Single Agent Makes a Difference." *Journal of Economic Dynamics and Control* 27:1917–38.

Ortony, Andrew, ed. 1993. *Metaphor and Thought*. 2nd ed. New York: Cambridge University Press.

Palumbo, Donald E. 2002. *Chaos Theory, Asimov's Foundations and Robots, and Herbert's "Dune": The Fractal Aesthetic of Epic Science Fiction*. Westport, CT: Greenwood Press.

Paulson, William. 1991. "Literature, Complexity, Interdisciplinarity." In *Chaos and Order: Complex Dynamics in Literature and Science*, edited by N. Katherine Hayles, 37–53. Chicago: University of Chicago Press.

Pepinsky, Hal. 1997. "Geometric Forms of Violence." In *Chaos, Criminology, and Social Justice: The New Orderly (Dis)Order*, edited by D. Milovanovic, 97–109. Westport, CT: Praeger.

Perelman, Chaim, and Lucie Olbrechts-Tyteca. 1969. *The New Rhetoric: A Treatise on Argumentation*. Translated by J. Wilkinson and P. Weaver. Notre Dame, IN: University of Notre Dame Press.

Piccardi, Carlo, and Simone Lazzaris. 1998. "Vaccination Policies for Chaos Reduction in Childhood Epidemics." *IEEE Transactions on Biomedical Engineering* 45:591–95.

Pickering, Andy. 1995. "After Representation: Science Studies in the Performative Idiom." In *PSA 1994: Proceedings of the 1994 Biennial Meeting of the Philosophy of Science Association*, 2:413–19. East Lansing, MI: Philosophy of Science Association.

Pigliucci, Massimo. 2000. "Chaos and Complexity: Should We Be Skeptical?" *Skeptic* 8:62–70.

Pitta, Julie. 1998. "Putting Out Feelers." *Forbes* 161:206–8.

Pomerance, Leon. 1971. "The Need for Guidelines in Interdisciplinary Meetings." *American Journal of Archaeology* 75:429.

Porter, Elise. 1991. "The Player and the Dice: Physics and Critical Legal Theory." *Ohio State Law Journal* 52:1571–97.

Porter, Theodore. 1994. "Rigor and Practicality: Rival Ideals of Quantification in Nineteenth-Century Economics." In *Natural Images in Economic Thought: "Markets Read in Tooth and Claw,"* edited by Philip Mirowski, 128–70. New York: Cambridge University Press.

Porush, David. 1991. "Fiction as Dissipative Structures: Prigogine's Theory and Postmodernism's Roadshow." In *Chaos and Order: Complex Dynamics in Literature and Science,* edited by N. Katherine Hayles, 54–84. Chicago: University of Chicago Press.

Posner, Richard A. 1972. "Volume One of the *Journal of Legal Studies*—An Afterword." *Journal of Legal Studies* 1:437–40.

Post, David G., and Michael B. Eisen. 2000. "How Long Is the Coastline of the Law? Thoughts on the Fractal Nature of Legal Systems." *Journal of Legal Studies* 29:545–84.

Priest, George L. 1981. "The New Scientism in Legal Scholarship: A Comment on Clark and Posner." *Yale Law Journal* 90:1284–95.

Prigogine, Ilya. 1993. "Bounded Rationality: From Dynamical Systems to Socio-Economic Models." In *Nonlinear Dynamics and Evolutionary Economics,* edited by Richard H. Day and Ping Chen, 3–13. New York: Oxford University Press.

Putnam, Hilary. 1990. *Realism with a Human Face.* Edited by James Conant. Cambridge: Harvard University Press.

———. 2002. *The Collapse of the Fact/Value Dichotomy and Other Essays.* Cambridge: Harvard University Press.

Pylyshyn, Zenon W. 1993. "Metaphorical Imprecision and the 'Top-Down' Research Strategy." In *Metaphor and Thought,* 2nd ed., edited by Andrew Ortony, 543–58. New York: Cambridge University Press.

Pynchon, Thomas. 1966. *The Crying of Lot 49.* New York: Harper and Row.

———. 1973. *Gravity's Rainbow.* New York: Viking Press.

Quine, W. V. O. 1994. "Epistemology Naturalized." In *Naturalizing Epistemology,* 2nd ed., edited by Hilary Kornblith, 15–32. Cambridge: MIT Press.

Rabe, David. 1985. *Hurlyburly.* New York: Grove.

Radzicki, Michael J. 1988. "A Note on Kelsey's 'The Economics of Chaos or the Chaos of Economics.'" *Oxford Economic Papers,* n.s., 40:692–93.

———. 1990. "Institutional Dynamics, Deterministic Chaos, and Self-Organizing Systems." *Journal of Economic Issues* 24:57–102.

Reichenbach, Hans. 1938. *Experience and Prediction: An Analysis of the Foundations and the Structure of Knowledge.* Chicago, University of Chicago Press.

Reynolds, Glenn Harlan. 1991. "Chaos and the Court." *Columbia Law Review* 91:110–17.

Rice, Thomas Jackson. 1997. *Joyce, Chaos, and Complexity*. Urbana: University of Illinois Press.

Richard, Claude. 1983. "Causality and Mimesis." *SubStance* 40:84–93.

Richards, R. A. 2001. "A New Aesthetic for Environmental Awareness: Chaos Theory, the Beauty of Nature, and Our Broader Humanistic Identity." *Journal of Humanistic Psychology* 41:59–95.

Roe, Mark J. 1996. "Chaos and Evolution in Law and Economics." *Harvard Law Review* 109:641–68.

Rorty, Richard. 1991. *Objectivity, Relativism, and Truth: Philosophical Papers*. Vol. 1. New York: Cambridge University Press.

Rosenberg, Alexander. 1994. "Does Evolutionary Theory Give Comfort or Inspiration to Economics?" In *Natural Images in Economic Thought: "Markets Read in Tooth and Claw,"* edited by Philip Mirowski, 384–407. New York: Cambridge University Press.

Rosser, J. Barkley, Jr. 1991. *From Catastrophe to Chaos: A General Theory of Economic Discontinuities*. Boston: Kluwer Academic Publishers.

Rostow, W. W. 1993. "Nonlinear Dynamics and Economics: A Historian's Perspective." In *Nonlinear Dynamics and Evolutionary Economics*, edited by Richard H. Day and Ping Chen, 15–17. New York: Oxford University Press.

Rouse, Joseph. 1987. *Knowledge and Power: Toward a Political Philosophy of Science*. Ithaca, NY: Cornell University Press.

Rudd, Jeffrey. 2005. "J. B. Ruhl's 'Law-and-Society System': Burying Norms and Democracy under Complexity Theory's Foundation." *William and Mary Environmental Law and Policy Review* 29:551–632.

Ruelle, David. 1991. *Chance and Chaos*. Princeton: Princeton University Press.

———. 1994. "Where Can One Hope to Profitably Apply the Ideas of Chaos?" *Physics Today*, July.

Ruhl, J. B. 1996. "Complexity Theory as a Paradigm for the Dynamical Law-and-Society System: A Wake-up Call for Legal Reductionism and the Modern Administrative State." *Duke Law Journal* 45:849–928.

Ruse, Michael, and Edward O. Wilson. 1986. "Moral Philosophy as Applied Science." *Philosophy* 61:173–92.

Russell, Robert John, Nancey Murphy, and Arthur R. Peacocke, eds. 1995. *Chaos and Complexity: Scientific Perspectives on Divine Action*. Vatican City State: Vatican Observatory Publications; and Berkeley: Center for Theology and the Natural Sciences.

Samuelson, Paul A. 1980. *Economics*. 11th ed. New York: McGraw-Hill.

———. 1983. "Rigorous Observational Positivism: Klein's Envelope Aggregation; Thermodynamics and Economic Isomorphism." In *Global Econometrics: Essays in Honor of Lawrence R. Klein*, edited by F. Gerard Abrams and Bert G. Hickman, 1–38. Cambridge: MIT Press.

Saperstein, Alvin M. 1984. "Chaos—A Model for the Outbreak of War." *Nature* 390:303–5.

Sato, Yuzuru, Eizo Akiyama, and J. Doyne Farmer. 2002. "Chaos in Learning a Simple Two-Person Game." *Proceedings of the National Academy of Sciences of the United States* 99:4748–51.

Savage, Dan. 2004. "Savage Love." *City Pages*, June 23.

Schabas, Margaret. 1994. "The Greyhound and the Mastiff: Darwinian Themes in Mill and Marshall." In *Natural Images in Economic Thought: "Markets Read in Tooth and Claw,"* edited by Philip Mirowski, 322–35. New York: Cambridge University Press.

Scheinkman, José. 1990. "Nonlinearities in Economic Dynamics." *Economics Journal* 100 (suppl.): 33–47.

Schenker, H. 1973. *Free Composition*. Translated by E. Oster. New York: Longman.

Schön, Donald A. 1993. "Generative Metaphor: A Perspective on Problem-Setting in Social Policy." In *Metaphor and Thought*, 2nd ed., edited by Andrew Ortony, 137–63. New York: Cambridge University Press.

Scott, Robert E. 1993. "Chaos Theory and the Justice Paradox." *William and Mary Law Review* 35:329–51.

Scott, Robert L. 1967. "On Viewing Rhetoric as Epistemic." *Central States Speech Journal* 18:9–17.

Searle, John R. 1993. "Metaphor." In *Metaphor and Thought*, 2nd ed., edited by Andrew Ortony, 83–111. New York: Cambridge University Press.

Sen, Amartya. 1987. *On Ethics and Economics*. Worcester, UK: Basil Blackwell.

Serletis, Apostolos, and Mototsugu Shintani. 2006. "Chaotic Monetary Dynamics with Confidence." *Journal of Macroeconomics* 28:228–52.

Serres, Michel. 1974. "Les sciences." In *Faire de l'histoire: Nouvelles approches*, vol. 2, edited by Jacques Le Goff and Pierre Nora. Paris: Gallimard.

Seuss, Dr. 1954. *Horton Hears a Who!* New York: Random House.

Sewell, Susan, S. Stansell, D. Lee, and S. Below. 1996. "Using Chaos Measures to Examine International Capital Market Integration." *Applied Financial Economics* 6:91–101.

Shaviro, Daniel. 1990. "Beyond Public Choice and Public Interest: A Study of the Legislative Process as Illustrated by Tax Legislation in the 1980s." *University of Pennsylvania Law Review* 139:1–124.

Shaw, R. 1981. "Strange Attractors, Chaotic Behavior, and Information Flow." *Zeitschrift für Naturforschung* 36a:80–112.

Shearer, Rhonda Roland. 1992. "Chaos Theory and Fractal Geometry: Their Potential Impact on the Future of Art." *Leonardo* 25:143–52.

Sherif, Muzaher, and Carolyn Sherif, eds. 1969. *Interdisciplinary Relationships in the Social Sciences*. Chicago: Aldine Publishing.

Sherwin, Richard K. 1994. "The Narrative Construction of Legal Reality." *Vermont Law Review* 18:681–719.

Shintani, Mototsugu, and Oliver Linton. 2003. "Is There Chaos in the World Economy? A Nonparametric Test Using Consistent Standard Errors." *International Economic Review* 44:331–58.

———. 2004. "Nonparametric Neural Network Estimation of Lyapunov Exponents and a Direct Test for Chaos." *Journal of Econometrics* 120:1–33.

Shone, Ronald. 1997. *Economic Dynamics: Phase Diagrams and Their Economic Application*. Cambridge: Cambridge University Press.

Short, Larry. 1991. "The Aesthetic Value of Fractal Images." *British Journal of Aesthetics* 21:342–55.

Shusterman, Ronald. 1998. "Ravens and Writing-Desks: Sokal and the Two Cultures." *Philosophy and Literature* 22:119–35.

Simons, Herbert W. 1990. "The Rhetoric of Inquiry as an Intellectual Movement." In *The Rhetorical Turn: Invention and Persuasion in the Conduct of Inquiry*, edited by Herbert W. Simons, 1–31. Chicago: University of Chicago Press.

Sinclair, M. B. W. 1997. "Statutory Reasoning." *Drake Law Review* 46:299–382.

Skarda, C., and W. Freeman. 1991. "How Brains Make Chaos in Order to Make Sense of the World." *Behavioral and Brain Sciences* 10:161–95.

Slethaug, Gordon E. 2000. *Beautiful Chaos: Chaos Theory and Metachaotics in Recent American Fiction.* Albany: SUNY Press.

Smith, Joseph H., and William Kerrigan, eds. 1984. *Taking Chances: Derrida, Psychoanalysis, and Literature.* Baltimore: Johns Hopkins University Press.

Smith, Peter. 1991. "The Butterfly Effect." *Proceedings of the Aristotelian Society* 91:247–67.

———. 1998. *Explaining Chaos.* Cambridge: Cambridge University Press.

Smith, Trig R. 2000. "The S.E.C. and Regulation of Foreign Private Issuers: Another Missed Opportunity at Meaningful Regulatory Change." *Brooklyn Journal of International Law* 26:765–95.

Sokal Alan. 1996. "Transgressing the Boundaries: Toward a Transformative Hermeneutics of Quantum Gravity." *Social Text* 46/47:217–52.

Sokal, Alan, and Jean Bricmont. 1998. *Fashionable Nonsense: Postmodern Intellectuals' Abuse of Science.* New York: Picador.

Sorger, Gerhard. 1998. "Imperfect Foresight and Chaos: An Example of a Self-Fulfilling Mistake." *Journal of Economic Behavior and Organization* 33:363–83.

Sorrentino, Gilbert. 1987. *Rose Theater.* Elmwood Park, IL: Dalkey Archive Press.

———. 1997. *Pack of Lies.* Normal, IL: Dalkey Archive Press.

Sternberg, Robert J., Roger Tourangeau, and Georgia Nigro. 1993. "Metaphor, Induction, and Social Policy: The Convergence of Macroscopic and Microscopic Views." In *Metaphor and Thought*, 2nd ed., edited by Andrew Ortony, 277–303. New York: Cambridge University Press.

Stewart, Ian. 1989. "Portraits of Chaos." *New Scientist,* November 4.

Stoicheff, Peter. 1991. "The Chaos of Metafiction." In *Chaos and Order: Complex Dynamics in Literature and Science,* edited by N. Katherine Hayles, 85–99. Chicago: University of Chicago Press.

Stoppard, Tom. 1993. *Arcadia.* London: Faber and Faber.

Stump, David. 1992. "Naturalized Philosophy of Science with a Plurality of Methods." *Philosophy of Science* 59:456–60.

Szegö, Giorgio P. 1982. *New Quantitative Techniques for Economic Analysis.* New York: Academic Press.

Taylor, Mark C. 1993. *Nots.* Chicago: University of Chicago Press.

Taylor, Richard P., Adam P. Micolich, and David Jonas. 1999. "Fractal Analysis of Pollock's Drip Paintings." *Nature* 399:422.

Thom, René. 1983a. "By Way of Conclusion." Translated by Robert E. Chumbley. *SubStance* 40:78–83.
———. 1983b. "Stop Chance! Silence the Noise!" Translated by Robert E. Chumbley. *SubStance* 40:11–21.
Toulmin, Stephen Edelston. 1958. *The Uses of Argument*. Cambridge: Cambridge University Press.
———. 1972. *Human Understanding*. Vol. 1. Princeton: Princeton University Press.
Tribe, Laurence H. 1989. "The Curvature of Constitutional Space: What Lawyers Can Learn from Modern Physics." *Harvard Law Review* 103: 1–39.
Trippi, Laura. 1989. "Fractured Fairy Tales, Chaotic Regimes." In *Strange Attractors: Signs of Chaos*. Catalogue organized by Alice Yang. New York: New Museum of Contemporary Art.
Tsuda, Ichiro. 2001. "Toward an Interpretation of Dynamic Neural Activity in Terms of Chaotic Dynamical Systems." *Behavioral and Brain Sciences* 24:793–810.
Vanden Heuvel, Michael. 1993. "The Politics of the Paradigm: A Case Study in Chaos Theory." *New Theatre Quarterly* 92:255–66.
van Eemeren, Frans H., Rob Grootendorst, Francisca Snoek Henkemans, J. Anthony Blair, Ralph H. Johnson, Erik C. W. Krabbe, Christian Plantin, Douglas N. Walton, Charles A. Willard, John Woods, and David Zarefsky. 1996. *Fundamentals of Argumentation Theory: A Handbook of Historical Backgrounds and Contemporary Developments*. Mahwah, NJ: Lawrence Erlbaum Associates.
van Fraasen, Bas C. 2002. *The Empirical Stance*. New Haven: Yale University Press.
van Peer, Willie. 1997. "Sense and Nonsense of Chaos Theory in Literary Studies." In *The Third Culture: Literature and Science*, edited by Elinor S. Shaffer, 40–48. Berlin: Walter de Gruyter.
van Staveren, Irene. 1999. "Chaos Theory and Institutional Economics: Metaphor or Model?" *Journal of Economic Issues* 33:141–67.
Varian, H. R. 1991. Review of *More Heat Than Light: Economics as Social Physics, Physics as Nature's Economics*, by Philip Mirowski. *Journal of Economic Literature* 29:595–96.
Veilleux, John. 1987. "The Scientific Model in Law." *Georgetown Law Journal* 75:1967–2003.
Walker, D. A. 1991. "Economics as Social Physics." *Economic Journal* 101:615–31.
Walton, Douglas. 1997. *Appeal to Expert Opinion*. University Park: Pennsylvania State University Press.
Weinberg, Steven. 1996. "Sokal's Hoax." *New York Review of Books*, August 8.
Weingart, Peter, and Sabine Maasen. 1997. "The Order of Meaning: The Career of Chaos as a Metaphor." *Configurations* 5:463–520.
Weintraub, E. R. 1991. *Stabilizing Dynamics: Constructing Economic Knowledge*. New York: Cambridge University Press.

Weissert, Thomas P. 1991. "Representation and Bifurcation: Borges's Garden of Chaos Dynamics." In *Chaos and Order: Complex Dynamics in Literature and Science*, edited by N. Katherine Hayles, 223–43. Chicago: University of Chicago Press.

Weisskopf, W. A. 1983. "Reflections on Uncertainty in Economics." *Geneva Papers on Risk and Insurance* 9:335–60.

Welchman, Jennifer. 1995. *Dewey's Ethical Thought*. Ithaca, NY: Cornell University Press.

Wertheimer, R., and M. Zinga. 1998. "Applying Chaos Theory to School Reform." *Internet Research: Electronic Networking Applications and Policy* 8:101–14.

Wheatley, Margaret J. 1992. *Leadership and the New Science: Learning about Organization from an Orderly Universe*. San Francisco: Berrett-Koehler.

White, James Boyd. 1984. *When Words Lose Their Meaning: Constitutions and Reconstitutions of Language, Character, and Community*. Chicago: University of Chicago Press.

Wildermuth, Mark. 1999. "The Edge of Chaos: Structural Conspiracy and Epistemology in *The X-Files*." *Journal of Popular Film and Television* 26: 146–57.

Willard, Charles A. 1983. *Argumentation and the Social Grounds of Knowledge*. University: University of Alabama Press.

Williams, Bernard. 1985. *Ethics and the Limits of Philosophy*. Cambridge: Harvard University Press.

Williams, Christopher R., and Bruce A. Arrigo. 2002. *Law, Psychology, and Justice: Chaos Theory and the New (Dis)order*. Albany: SUNY Press.

Williams, Joan C. 1987. "Critical Legal Studies: The Death of Transcendence and the Rise of the New Langdells." *New York University Law Review* 62:429–96.

Wilson, David Sloan, and Elliot Sober. 1998. *Unto Others: The Evolution and Psychology of Unselfish Behavior*. Cambridge: Harvard University Press.

Wilson, Edward O. 1980. *Sociobiology: The Abridged Edition*. Cambridge: Harvard University Press.

Wise, M. Norton. 1993. "Mediations: Enlightenment Balancing Acts, or the Technologies of Rationalism." In *World Changes*, edited by Paul Horwich, 207–56. Cambridge: MIT Press.

Wise, Michael O. 1995. "Antitrust's Newest 'New Learning' Returns the Law to Its Roots: Chaos and Adaptation as New Metaphors for Competition Policy." *Antitrust Bulletin* (Winter): 713–77.

Wolfson, Nicholas. 1989. "Efficient Markets, Hubris, Chaos, Legal Scholarship and Takeovers." *St. John's Law Review* 63:511–36.

Woodford, Michael. 1987. "Equilibrium Models of Endogenous Fluctuations: Cycles, Chaos, Indeterminacy, and Sunspots." Department of Economics, University of Chicago. Mimeo.

———. 1989. "Imperfect Financial Intermediation and Complex Dynamics." In *Economic Complexity: Chaos, Sunspots, Bubbles, and Nonlinearity*, edited by William A. Barnett, John Geweke, and Karl Shell, 309–34. Cambridge: Cambridge University Press.

Wylie, Alison. 1995. "Discourse, Practice, Context: From HPS to Interdisciplinary Science Studies." In *PSA 1994: Proceedings of the 1994 Biennial Meeting of the Philosophy of Science Association,* edited by David Hull, Micky Forbes, and Richard M. Burian, 2:393–95. East Lansing, MI: Philosophy of Science Association.

Young, T. R. 1997. "The ABCs of Crime: Attractors, Bifurcations, and Chaotic Dynamics." In *Chaos, Criminology, and Social Justice: The New Orderly (Dis)Order,* edited by D. Milovanovic, 77–96. Westport, CT: Praeger.

Zants, Emily. 1996. *Chaos Theory, Complexity, Cinema, and the Evolution of the French Novel.* Lewiston, ME: Edwin Mellen.

Zhou, Y., L. Ma, and L. Wang. 2002. "Chaotic Dynamics of the Flood Series in the Huaihe River Basin for the Last 500 Years." *Journal of Hydrology* 258:100–110.

Index

ab